TIME TO UPDATE

7 Areas to Integrate Digital Discipleship Into Your Children's Ministry Strategy

Brittany Nelson

DEEPER KIDMIN
growing kids deeper

Foreword by Esther Moreno

Published by Deeper KidMin | deeperkidmin.com

Book Cover by Ajibola O.

Graphics and Illustrations by Brittany Nelson

1st edition 2023

ISBN (print): 979-8-9886743-0-6

ISBN (ebook): 979-8-9886743-2-0

ISBN (audio): 979-8-9886743-5-1

"Brittany Nelson is a young kidmin leader that I've been watching and learning from for some time! What drew me to her was her passion to help others be the "disciplers" Jesus has called them to be. This new book *Time To Update* is a must-read for anyone who has said "Technology scares me" or if you still use a paper calendar! (Boomers order two copies and give one to a friend.) Believe me, friend, this digital world is not going away, and thank the Lord Brittany has provided such a practical resource to help kidmin leaders of all ages navigate what digital discipleship should look like for future generations. This is my new favorite Kidmin book!"

Jim Wideman

Kidmin Pioneer, Pastor, Author, Leadership Coach

"*Time to Update* is a game-changer for anyone involved in children's ministry. With clear and practical guidance, Brittany explores the impact of digital discipleship on key areas such as safety, teaching, partnering with parents, and more. This book is a must-read for anyone seeking to navigate the intersection of faith, technology, and children's ministry effectively."

Ryan Frank

KidzMatter CEO & Publisher

"In children's ministry the word "digital" is typically reserved for check-in processes and curriculum delivery, but to ignore the endless possibilities for digital discipleship is a disservice to the families we love and serve. In *Time to Update*, Brittany not only builds a case for prioritizing your digital strategy, but also provides practical and easy-to-understand insight. This is a must-read for every children's minister."

Jessica Bealer

Director of Family Ministry Services, Generis

"In her book, *Time to Update*, Brittany Nelson lives up to her Deeper KidMin brand through her thoughtful approach to discussing the important role of digital technology in a local, child discipleship ecosystem. Her strategic overview of several key areas will give your team much to discuss, consider, and wrestle through together. The combination of her local church experience, her bias toward action, her creative ideas, and her ability to think deeply make this a must-read as you disciple children in today's secular age."

Matt Markins

Awana, President and CEO

Child Discipleship Forum, Co-Founder

"*Time to Update* is a relevant and real look at what is required to reach today's kids and families. It is not only a practical guide, but also a heartfelt come-alongside encouragement for today's Kids Ministry Leader. Whether you are just dipping your toe into the digital world, or a deep diver, Brittany's personal anecdotes, thoughtful coaching, and tangible next-step tools will equip you effectively for digital discipleship."

Christie Penner Worden

Chief Mission Officer, RaiseUpFaith

"There's no way around it, this book is a MUST-READ for anyone who works with kids. It's the most beneficial book that I've read in years and has given me a glimpse behind the curtain of how today's kids think and react in this digital world. *Time to Update* provides tons of small changes you can make immediately, but also outlines how to revamp every aspect of your ministry. If you truly want to reach today's kids, be willing to set aside your old way of thinking and enter the only world our kids have known... and that is the digital world."

Tina Houser

Writer, speaker, editor, children's pastor, & Silly Grandma

"Finally! A book with simple, doable strategies to engage today's kids in a gospel that's still alive. Thank you, Brittany, for writing to the everyday lay leader, Sunday School teacher, and children's director. Relatable, direct, and practical, this book is essential reading for anyone who cares deeply about reaching and teaching the upcoming generation."

Laurie Acker
Founder of Small Church Ministry

"Do you hear the words "digital discipleship" and immediately feel overwhelmed? Time to Update is the perfect book to help you realize the importance of using technology and the digital world to reach your church families. Brittany does a phenomenal job of breaking down what digital discipleship is and how to implement it into your children's ministry. She equips you with knowledge and a specific plan of what to do. This is a great book for beginners and seasoned Kidmin leaders!"

Vanessa Myers
Children's Ministry Director, Dahlonega, GA
Founder/Creator of Family Faith Builders, Author

"As a ministry leader, I know how important digital discipleship is - but I also know that it often feels overwhelming - and adds one more thing to my to-do list each week! If you've felt (or feel!) the same way, Brittany's new book is for you. *Time to Update* is filled with insightful ideas and SO MUCH practical information - it left me feeling empowered and equipped to create a digital discipleship strategy for my ministry. I can't wait to share about this book with others!"

Julia Ball
Next Gen Pastor
Owner of EquipFamily.com

"My favorite part of this great ministry read is the fact that Brittany doesn't want leaders to reinvent the wheel or throw out everything they are doing, she wants to provide important resources and discourse on digital natives and how we can evolve ministry to engage and lead kids in an ever-changing society. From the start, she provides specific information that can be implemented day one, as well as ideas that will have you thinking and growing. This isn't a one-time read, but a new tool in your ministry library that you will reach for again and again as you strive to passionately, and relevantly, reach children for His kingdom."

Jolie Sims
Family Pastor
Celebration Church, Hoschton, GA

"Digital Discipleship! Wow! Brittany takes this concept that seems intimidating and overwhelming and breaks it down for the children's ministry leader. The digital age is not going anywhere, and *Time to Update*, shows you how to take advantage of the tools right in front of you to expand your reach and partner with parents in a more effective way."

Debbie Rhoads
Children's Ministry Director
Grace Chapel, Franklin, TN

"Digital Ministry is how we are going to reach this generation and beyond for Jesus. Brittany does a wonderful job capturing the practical steps to make that happen. From developing a strategy to putting that strategy into practice, this is a definitive guide for digital discipleship in your Kid's Ministry. A must-have in your toolbox to minister to kids!"

Carl Barnhill
Owner, Church Visuals

"There is nobody I trust more to give a detailed strategy for digital discipleship within children's ministry than Brittany Nelson. Brittany has built an incredible ministry through consistent love and care shown through the Deeper KidMin daily social media posts, and weekly newsletter and blog. Brittany outlines the importance of digital discipleship within the context of today's digital natives, their parents, and kidmin volunteers. If you follow Brittany's advice and the action steps in *Time to Update,* you are sure to enhance your ministry to children and their families."

Katy Herrera-Fonseca
Creator of MinistryArk.com

"Technology integration in ministry settings is far more than a conversation about what app to use or how best to show videos in Sunday school. *Time To Update* will provide the gentle push you need to consider new practices in ministry: not just good digital practices, but good educational practices as well. *Time to Update* provides doable ideas and forward-thinking encouragement to consider how your Children's Ministry can thrive in today's reality of both digital and in-person ministry."

Christie Thomas
Former Children's Ministry Director
Founder of the online family ministry, *Little Shoots, Deep Roots*

"Brittany has lived it. And now she shares what she has lived. When it comes to including 'digital' in your 'discipleship' the question can't be 'If?" It must be "How?" In *Time to Update,* Brittany combines theory, inspiration, and application in a way that you feel encouraged and guided by a wise friend. If you want to help the next generation draw closer to Jesus, Brittany's book will help you do it!"

Keith Ferrin
Speaker and Author of *How to Enjoy Reading Your Bible*

"I am grateful to call Brittany a friend first and ministry peer second. I have watched her focus on the writing of this book for over a year, which, to me, displays a passion for the children taught, the leaders trained, and Her Savior. *Time to Update* will help church leaders to honestly evaluate their ministries and intentionally use technology. Many parents will check out a church online before they ever step into a building, so having an updated website showcasing events, safety procedures, and schedules is important. And that is just one simple way we can use technology! This book will hold the hands of ministry leaders who are nervous about technology, but understand that digital natives, parents, and volunteers are worth the effort needed to learn, and implement, new discipleship methods and strategies. This is a must-have book for anyone ministering to children and families."

Anne Marie Gosnell

Author of the *Bible Object Lessons for Kids* series

Owner of FutureFlyingSausers Resources and the Bible Creative Online Conference

DOWNLOAD THE AUDIOBOOK FREE!

READ THIS FIRST

To say thanks for downloading or purchasing my book, I would like to give you the Audiobook version 100% FREE!

I know you're more likely to finish the book if you have the audiobook too, so you can read while cutting out crafts for Sunday, organizing the supply closet, or driving to and from mid-week programming. I even narrated the book myself so it will feel like we are having a conversation.

Instead of paying $10-$20 for the audiobook,
I'd like you to have it for free:

deeperkidmin.com/timetoupdateaudio

This book is dedicated to
my girls, Holloway and Kinsley,
to the children of their generation,
and to the leaders helping these digital natives
grow deeper in their faith.

Contents

Foreword

DEAR READER,

If you are holding this book in your hands, then you recognize the profound importance of this moment in children's ministry. It is no question that the global pandemic posed a host of challenges to churches across the country, and children's ministries were no exception. Whether we wanted it or not, we were confronted with the reality that we had to do something to reach children and families who were quarantined in their homes. With varying degrees of success, many of us rose to the occasion and began to implement digital strategies. Unfortunately, when the pandemic began to subside, many of us abandoned those digital strategies without a second thought.

Time To Update is a clarion call for leaders to embrace the importance of a healthy digital strategy. In a world populated by digital natives, we can no longer ignore or avoid the importance of a robust digital discipleship strategy. From developing a digital strategy to leveraging technology in administration, Brittany masterfully covers a spectrum of ways leaders can harness technology in various ministry areas.

If you are someone who gets overwhelmed at the thought of anything virtual, digital, or technology-based, rest assured, this book is for you. Brittany provides practical and easy-to-use steps for even the most novice leader. Conversely, if you have a mature digital strategy, this book will challenge you to take your ministry to the next level. I am so grateful Brittany has produced this comprehensive and practical resource for the children's ministry community. No question, this book is a resource every children's ministry leader should have in their library.

Brittany helps us to dive headfirst into these digital waters, but does not leave us to drown. Rather, with practical examples and well-researched techniques, she guides us every step of the way with effective strategies guaranteed to help you reach this generation of digital natives. Brittany is a brilliant voice in the children's ministry world and I am blessed to call her friend. Her passion for the next generation of children and families is clear and resonates on every page. It is my sincere desire, as you read this book, you do so with full appreciation that the investment you're making will have an eternal impact.

Esther Moreno
Founder of Child's Heart LLC

Introduction

WE UPDATE EVERYTHING IN our lives and ministries. We regularly update our phones and computers to keep them running correctly. We update the worship songs we use and the curriculum we teach. Even when transitioning from one teaching series to another, we often update the physical environment of the children's ministry with new backdrops, décor, and imagery to stay relevant to the series theme. So, why wouldn't we also update our methods and strategy for discipleship to include the digital world in which everyone now spends a large portion of their time?

We live in an age of acceleration, meaning technology is accelerating at such an exponential rate that the human mind cannot possibly keep up or adapt in time. [1] While this shift to a digital world may seem staggering, consider the change from oral to written language. Writing itself is a form of technology. When the Greeks invented an alphabetic system of just 24 characters around 700 BC, it was a significant advancement from pictographs and cave drawings, and it drastically changed how the world communicated.

Not everyone was in favor of it.[2] Socrates, one of the most famous philosophers in history, argued against the written language and thought it would weaken one's memory and communication skills. We know this, ironically, because Socrates' pupil, Plato, recorded Socrates' negative opinion of writing in the written word. But this new technology of the written word expanded the world in numerous ways, rushing communication and life into new eras of growth.

The COVID-19 pandemic sped up the digital world's impact on ministry and brought the necessity of digital discipleship to the forefront of ministry strategy. The pandemic didn't just allow children's ministry leaders to dip their toes into the waters of digital ministry. It pushed all of us off the cliff and straight into the deep end!

Now that the world is moving into post-pandemic life, leaders can resume pre-pandemic ministry practices, but *wise* leaders will keep exploring the approaches and tools they learned when digital ministry was the only option. In this new age, leaders must intentionally incorporate digital discipleship into children's ministry strategy.

What is Digital Discipleship?

In the broader sense, discipleship is the process of guiding someone into a deeper, more vibrant relationship with Jesus through teaching, encouragement, and personal connection. In her book *From Social Media to Social Ministry*, Nona Jones emphasizes the importance of relationships in the discipleship process. She defines discipleship as "the product of dialogue about how to apply [biblical] content to people's lives in a way that leads to continual transformation."[3]

Discipleship in children's ministry means partnering with parents in that continual process of growing kids deeper in their relationship with Jesus. There may be more hand motions in the worship songs and glitter in the supply closets, but the goal of discipleship remains the same.

2 Peter 3:18 says, *"Grow in the grace and knowledge of our Lord and Savior Jesus Christ. To Him be glory both now and forever!"* Discipling kids means helping them grow deeper in their relationship with and understanding of Christ and cheering them on as they go serve others in His name, for His glory. Ministry leaders want kids to know Jesus, understand their place in His story, and want to follow His example of a life of servitude.

In children's ministry, your strategy, plans, and efforts radiate from this desire to make disciples of the next generation. But to stay relevant to this next generation of kids and families, delivering the gospel message *must* include technology and a digital approach.

Digital discipleship is the process of making disciples through digital methods. Notice that the end goal, making disciples, stays the same. But the methods and procedures for doing so adopt a digital influence. The digital world offers children's ministry leaders another tool to share the gospel, and I'd argue it's the most powerful one. How leaders communicate and teach the gospel message must be updated to reach the digital natives they now serve.

When I say digital discipleship, I don't just mean live streaming your children's ministry service or using a downloadable curriculum. And I'm not advocating that you hit play on a Bible story video and call it a Sunday. A digital children's ministry, like a YouTube channel or weekly Zoom meetings, might be *part* of your overall digital discipleship strategy, but they aren't discipleship strategies in and of themselves.

These resources merely act as another tool or program you use to reach the kids and families you serve, both inside and outside your church building. A digital approach to children's ministry involves intentionally using technology to enhance and improve various areas within the ministry, from administrative tasks no one ever sees to communication with families to teaching Bible lessons on Sunday mornings.

When Jesus invited His disciples to follow Him, He told them: *"Come, follow me, and I will send you out to fish for people"* (Mark

4:19). Digital discipleship doesn't replace the church or replace the necessity of in-person relationships. It's not reeling in the fishing lines you have already cast out (the building, in-person programs, etc.). Digital discipleship is throwing out *another* fishing line, giving you *another* tool, and increasing opportunities to connect with kids and families.

When I talk with ministry leaders about the idea of digital discipleship, people tend to fall into two camps: 1) They're afraid of or annoyed with technology and want to ignore it, or 2) They're obsessed with technology, and it has muddied their mission.

To the first group, take a deep breath. You're not in this alone; we can figure it out together. To the second group, it's time to measure and evaluate the effectiveness of your tools so they stay within your purpose in ministry.

A balance between these mindsets requires two things:1) Exponential adaptability and 2) Acute discernment. You need exponential adaptability to help continually adjust to new conditions for the sake of the gospel and the next generation. You can't just cover your ears and plow forward as you always have, hoping the digital world will disappear. But you also need acute discernment to determine which elements of these new conditions are practical rather than allow yourself to get caught up in the latest release. Adopting open adaptability and focused discernment will allow you to minister to the digital world more effectively.

Three Reasons Ministry Leaders Can't Ignore Digital Discipleship

There are many positive and negative opinions about the next generation's obsession with the digital world and how the digital world is influencing development, relationships, and society. But, no matter your thoughts about the influence of technology, there are three reasons ministry leaders can't ignore digital discipleship:

First, digital is all the next generation knows. As a millennial, I'm a digital native, meaning I'm one of the generations who doesn't know life without the internet. We've spent our entire lives surrounded by computers, video games, cell phones, streaming services, and all the other tools and apps found in the digital world. The next generation of kids and families doesn't just use the internet; we live on it. Digital native generations (those born after the 1980s, though some argue that the first actual generation of digital natives was born after 1995) receive judgment for our misuse of social media and the digital world. I acknowledge the adverse effects this digital age can have on our relationships, attention spans, and self-esteem, but the digital world is the world into which we were born. Children's ministry leaders can fault a generation of kids and parents for living in the only reality they've ever known, or leaders can step into their realities to reach them with the gospel and teach them about Jesus. Using technology and the digital world helps leaders stay relevant to the kids and families they serve.

Second, digital provides opportunities for the church like never before. Digital media and social networking allow for faster information sharing in a way accessible to anyone with an internet connection, which is nearly 65% of the world's population as of January 2023.[4] People can communicate more quickly and easily; the digital world offers flexibility and quick adaptability. Ultimately, the internet and social media provide a mission field for the church that moves beyond the confines of physical space or location. Digital discipleship offers an extension of your physical ministry into the online world.

Finally, like it or not, the digital world is here to stay. Churches must embrace the tool of technology or risk becoming completely irrelevant to the next generation. According to a recent Barna study, *Six Questions About the Future of the Hybrid Church Experience*, only 42% of Millennials (the parents of the kids in your ministry) prefer in-person worship, meaning that the majority don't.[5] This same study also showed that parents are looking for hybrid (in-person and online) options for church services for their families. Kids will continue growing up as digital natives, knowing how to operate a tablet and scroll through a smartphone at a young age. Technologies will continue to advance, change, adapt, and improve faster and faster. For better or worse, our world's addiction to and reliance on technology means the digital world is a permanent fixture in our lives.

Let's Be Intentional

You've already implemented some aspects of digital discipleship and may not even realize it. Do you interact as your ministry on social media? Have a digital check-in station? Do you use videos at any point in your children's ministry programming? If you answered yes to any of these questions, give yourself a high five! You're already using digital technology to reach the next generation.

Time to Update will help you be more intentional so that digital discipleship becomes part of your ministry strategy rather than merely a coincidence. Unfortunately, there's no guidebook or easy formula to implement digital discipleship. The very nature of the digital world means life and ministry are constantly shifting, adapting to new technologies and information.

During the many interviews I conducted while researching this book, Jon Torres, a kidmin leader in Washington state, said, "I'm excited for when someone figures it out" – "it" being using digital discipleship effectively in children's ministry. [6]

While I don't claim to have it all figured out and never will, I have been experimenting, observing, researching, and conversing with ministry leaders in this area of digital discipleship for over two years to create the content of this book. Throughout each chapter, we'll take a robust look at the overall discipleship strategy and explore seven different areas to incorporate digital discipleship into every aspect of children's ministry.

Each chapter also includes a few practical action steps to help you immediately implement the ideas in your ministry.

Chapter 1 dives into what it looks like to develop and evaluate your current ministry strategy.

Chapters 2-8 explore seven different areas of children's ministry and how digital discipleship may affect and improve them.

Chapter 9 offers closing action steps to help you implement and move forward with your newly developed digital discipleship strategy.

Make sure to check out the Bonus Tools & Resources section at the end of the book for more tips and ideas.

Scan the QR code here or head to deeperkidmin.com/timetoupdateresources to access and print all the tools and resources mentioned in the book.

Time to Update will help you reach the next generation more effectively so that their lives glorify God online and in person. I pray that the tools, ideas, and strategies within these pages equip you to teach and show the love of Jesus Christ in a relevant, engaging way for the digital natives you serve now and in future generations.

Leader, it's time to update.

Chapter 1

Developing a Digital Ministry Strategy

HAVE YOU EVER PLAYED the board game Saint Petersburg? It's an obscure board game that's no longer in distribution, but it's one of my husband's all-time favorites. The goal of the game is to rack up victory points, which players earn through buying and collecting a variety of worker cards, building cards, and aristocrat cards throughout the game. My husband tried to share his strategies with me the first time we played the game together. At first, I felt confident about the gameplay and thought I had a fighting chance.

Then halfway through the game, I realized he was implementing a different strategy than the one he originally shared with me, and it was more effective! I lost the first game but was ready to employ the method he used next time. So for the next round, I adopted the strategy he beat me with the first time. Feeling confident in this new strategy, I saw victory close at hand until the last few rounds of the game when I noticed he was using yet another approach and would beat me by more than a little.

For the sake of our marriage, my husband and I don't play Saint Petersburg often, and even though I hate losing, I'm always astounded at the many strategies and approaches my husband can think up to help him win (every time). He says he adjusts the approach he uses based on the cards dealt. Like Saint Petersburg and other board games, leading a children's ministry well requires an intentional strategy flexible enough to adjust to the circumstances and context.

The global and eternal impact leaders can make on the next generation of kids and families is too significant *not* to approach ministry with an intentional game plan. A thriving children's ministry will have a specific strategy or action plan to achieve the overall aim or goal. You may have this strategy written down and detailed on paper, or it may live in your head. Prevalence of the digital world and the digital natives in your ministry requires adjustments to the overall strategy, yet building a digital discipleship strategy is much simpler than many leaders think.

Developing a digital ministry strategy does not mean you ditch what you have and start from scratch. It means you become more strategic and intentional with building digital tools and resources into your already-established strategy. Your digital ministry strategy becomes part of what you're already doing to disciple the next generation.

In his online course Lead Digital, Josh Denhart, founder of LeadMinistry.com, says, "Leading digitally helps *support* in-person ministry, not *supplant* in-person ministry."[7]

Tiffany Deluccia from the Unstuck Group (a consulting organization for strategic planning in churches) explains it best: "Digital strategies are going to become embedded parts of your ministry strategy. It won't be something separate. It won't be something different. They will be elemental, essential, components of how you do what you do to help people far from God meet Him and take next steps in their relationship with Him."[8]

So, take a deep breath. You won't be reworking your entire ministry strategy for digital discipleship. You will begin incorporating digital discipleship by clearly defining your current ministry strategy, evaluating where you are, and brainstorming ideas for the future.

Start with identifying your mission, purpose, and current ministry strategy. I suggest grabbing some paper or a notebook to jot down ideas and plans along the way.

Step 1: Identify Your Mission and Purpose

Does your ministry have a mission or vision statement? If so, take a moment to write it down and refresh your memory. If not, consider your church's vision statement and create a related but more specific one for your children's ministry.

This vision statement often influences your overall discipleship strategy. In the Introduction, we talked about the goal of discipleship, which is leading people into a deeper and more vibrant relationship with Jesus. Your discipleship strategy is the practical way(s) that goal of discipleship gets carried out. The vision statement helps you stay focused on the end goal and provides strategic specifics that answer questions like:

- Who are you trying to reach?
- What are you doing to reach them?
- Where are you trying to reach them?
- When are you trying to connect with them?
- How are you reaching them?

For example, in my first children's ministry, our vision statement was: "Through a culturally relevant but biblically reverent ministry, Rising Kids strives to partner with parents so that kids will know Jesus, understand their place in His story, and want to follow His example of a life of serving." In short, we used the phrase "Grow and Go." We wanted kids to grow deeper in their relationship with Jesus and then go out and serve others in His name.

Ultimately, the vision statement acts like a colander or pasta strainer for your ministry, allowing you to run all new ideas, events, and activities through that filter. One hundred percent of the ideas and suggestions you hear may be good ideas, but if they fall outside the overall vision of your ministry, then they aren't worth putting your time, energy, and effort into them. So, let your vision statement be a filter; pour all your ideas through them and see what sticks. Your vision statement builds the foundation for the rest of your ministry strategy, so it's vital that you take time to define it.

Discipleship Strategy

A discipleship strategy is different from a curriculum scope and sequence. Your curriculum may be part of your discipleship strategy, but your discipleship strategy is much more than your curriculum. Because discipleship is a lifelong process with many stages, many ministry leaders use a ministry funnel or discipleship pathway approach to create strategic plans for their ministry.

Ministry Funnel: In the ministry funnel model, there are four stages people tend to fall into: community, curious, committed, and commissioned. The funnel shape is helpful because the number of people in each phase tends to decrease as the stages progress.

The community stage at the top of the funnel refers to people who maybe just started coming to your church or they know of your church because they drive by it on their way to work. They may or may not know Jesus and aren't involved in a church body.

The curious stage includes people who have just started checking out your church and trying to decide if it's the right place for them and their families. They may also be new believers who are still curious about who Jesus is.

The committed stage includes the people who have made decisions to follow Christ and are committed members of your church. They may be volunteering in some areas, and you can count on them to attend your events and services.

The final stage, the commissioned step, includes your key leadership, volunteers, and people living out their faith daily.

Your ministry strategy should help move people down the funnel from one stage to the next. A healthy ministry strategy will provide events, services, programs, and intentionality for each of the four stages within the funnel. Churches tend to focus on one or two stages more than others, which is OK. But all churches should strive to offer something to people at each step in the funnel.

Recently, this ministry funnel model has come under criticism, as some have argued that the opposite approach is the true discipleship way of Jesus. Pete Scazerro, the author of *Emotionally Healthy Spirituality*, contends that the ministry funnel is ineffective and instead offers what he labels as a "Jesus Discipleship Approach" (see diagram below).[9] He encourages readers to pursue the funnel strategy on the right instead of the one on the left.

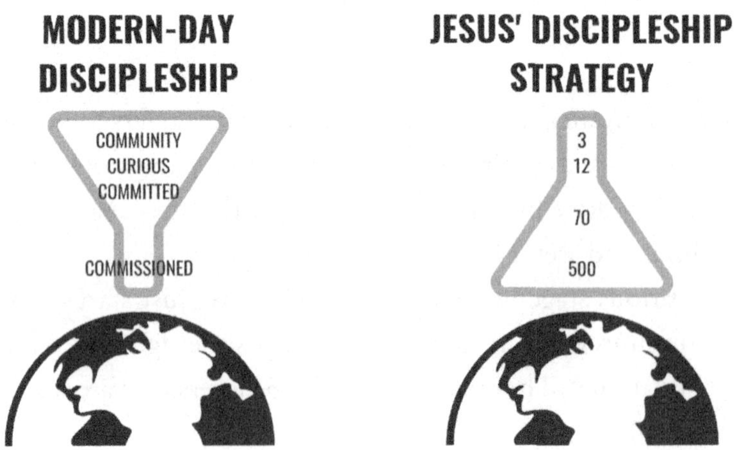

Rather than seeing these funnel models in opposition of each other, I propose they are a continuation of each other. So, rather than being opposite approaches, an effective ministry funnel on the left will lead to the discipleship funnel on the right. These two funnels are not separate or opposing strategies but different parts of the same design. This combined model looks like this:

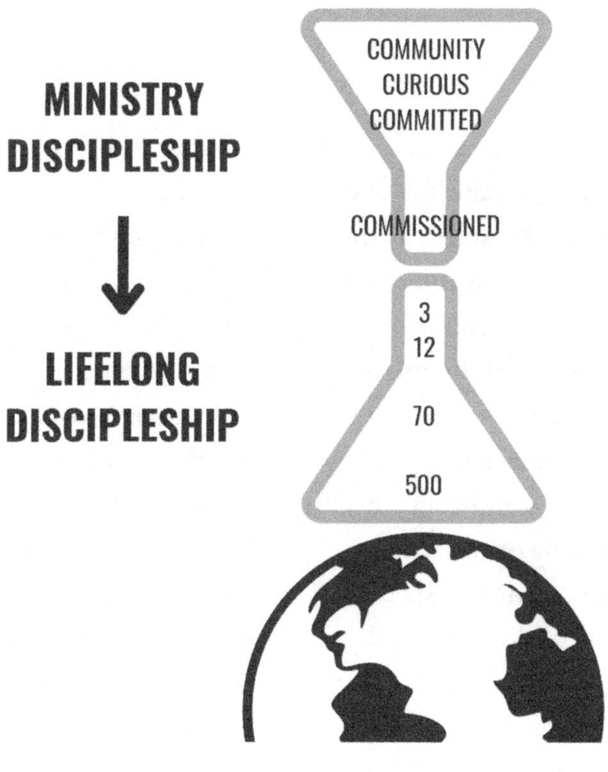

An effective ministry discipleship strategy will lead to an individual's lifelong discipleship strategy that parallels Jesus' discipleship. Rather than being mutually exclusive, the two processes are mutually dependent.

Discipleship Pathway: Another variation of Pete Scazzero's proposed "Jesus' Discipleship Strategy" is often found in a discipleship pathway. The discipleship pathway removes the funnel visual and positions discipleship as a journey or timeline.

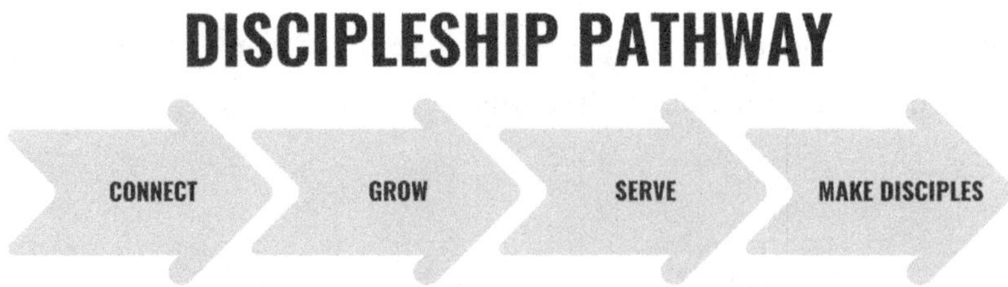

A quick Google search about discipleship pathways will offer several variations and examples. A discipleship pathway typically includes the same or similar stages as the ministry funnel, though they may have different names. Rather than focusing on groups or categories of people, the discipleship funnel focuses on an individual's progression of faith or personal discipleship. In Connect, the person will build connections with your church and God. Their relationship with Jesus will begin and deepen in the Grow stage, leading them to Serve within your church and community. Eventually, they'll move to Making Disciples and reaching others daily.

Which ministry strategy approach resonates most with your church's philosophy and your leadership? Can you identify what stage(s) of the funnel or pathway each of your events, programs, or initiatives reach?

Whichever approach or ministry model you choose, having a clear discipleship strategy is vital for any ministry. Incorporating digital discipleship into your plan means using digital technologies to help you carry out your vision more effectively. The remaining chapters look at different areas of children's ministry and explore ideas for incorporating digital discipleship into the overall vision.

Step 2: Evaluate Where You Are

Before deciding how to move forward with digital discipleship, you must know where you are now. Assess your ministry's current strengths, weaknesses, opportunities, and threats through a SWOT analysis. A SWOT analysis allows you to identify internal and external positives and negatives your ministry faces. SWOT can apply to various topics; we'll focus on digital discipleship. You can download a free printable of the SWOT chart from the Time to Update Resources page online at deeperkidmin.com/timetoupdateresources.

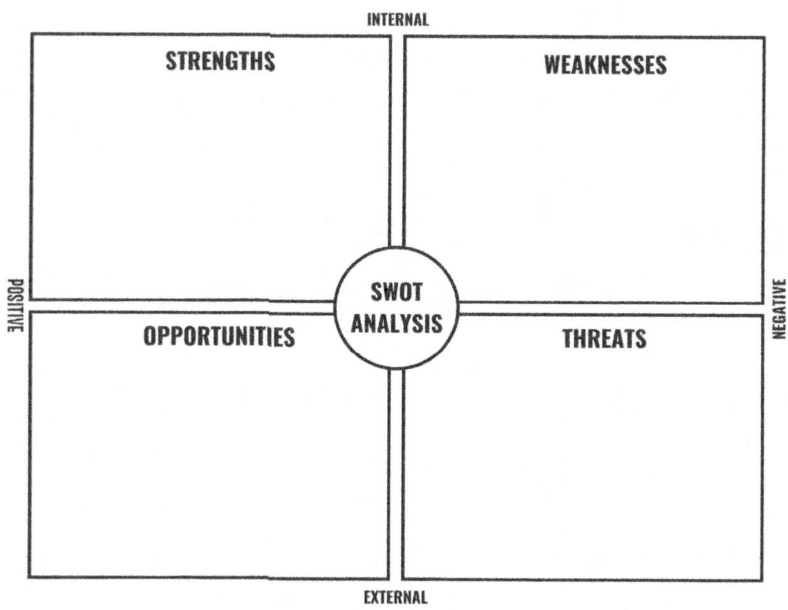

The SWOT analysis has four quadrants: strengths, weaknesses, opportunities, and threats.[10]

Strengths – Strengths are things your ministry is already doing well or the unique resources or people your church has access to. Strengths are internal factors, so you can build on them and use them to your advantage.

Weaknesses – Weaknesses are areas where your ministry could improve or where resources are needed. Weaknesses are also internal factors, so you can often address and overcome them.

Opportunities – Opportunities are external factors that you or your ministry could take advantage of. These could be new resources, emerging trends you could lean into, or any strengths you've yet to adopt into your strategy. Opportunities are external factors because they're beyond your control.

Threats - Threats are anything that could negatively impact your ministry from the outside or any obstacles your ministry currently faces. Like opportunities, threats are external factors and are often beyond your control.

Use the following questions and the example SWOT Chart to help you complete your own SWOT analysis of your digital discipleship strategy.

Questions to identify strengths:

- Where is our ministry succeeding when it comes to digital discipleship?
- What do we do well in or with the digital world?
- What have families told us they like about us?
- What's unique about our ministry and its digital discipleship strategy?
- What digital assets and resources do we utilize?

Questions to identify weaknesses:

- What can we improve when it comes to digital discipleship?
- What are our families dissatisfied with?
- Where do we need more digital knowledge or resources?

Questions to identify opportunities:

- What emerging digital trends can we take advantage of?
- What's happening in our community that might contribute to the growth of our ministry and digital discipleship approach?
- What external elements could we tap into to strengthen and support our digital discipleship strategy?

Questions to identify threats:

- What external elements cause our digital discipleship strategy trouble, frustration, or roadblocks?
- What changes may be coming (in the community or culturally) that require our church to adapt?

Example SWOT Chart:

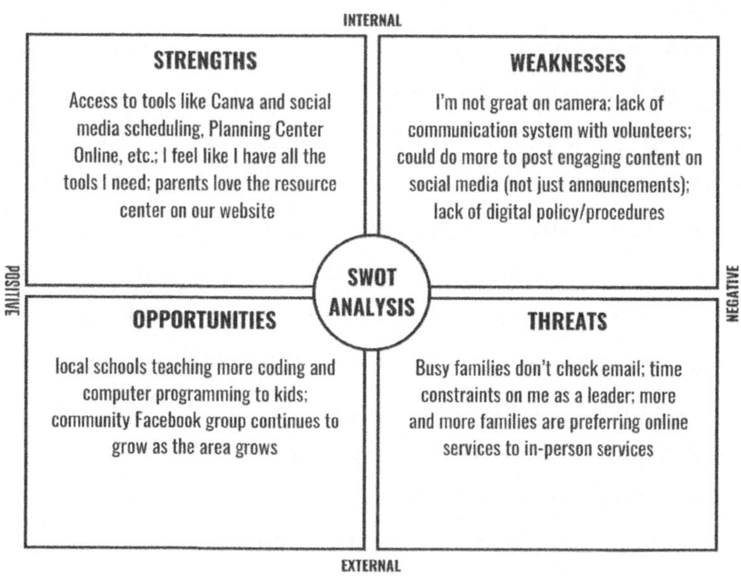

While you can easily conduct a SWOT analysis independently, it can also help to bring others into the conversation. Consider asking your senior pastor, fellow staff members, key volunteers, and parents to join you in evaluating your ministry and current digital discipleship strategy.

Take a break from reading to complete your SWOT chart regarding your digital discipleship efforts, then come back for the next step in the ministry strategy process.

Step 3: Dream Big and Brainstorm

Whew. Evaluation and analysis of your ministry or leadership can be challenging, but this next step in the strategic planning process is fun! Allow yourself to brainstorm, daydream, and explore new ideas and initiatives to help turn your threats into opportunities and your weaknesses into strengths. You won't be able to change some things, but knowing the threats and weaknesses will help you intentionally work around them.

Spend time in prayer, talking with God about what you want to see in your ministry and how He might further lead you in digital discipleship. Keep a running list close by as you continue reading this book in case any ideas stick out.

As you brainstorm, don't hold back. When you list ideas, defer judgment until later and prioritize quantity. How many ideas can you come up with to reach the families in your ministry through digital means? How many ways can you implement the tools found in this book to help you teach digital natives more effectively? When integrating digital discipleship into your strategy, what would be on your "dream list" of resources, people, and tools? Just let yourself brainstorm freely and write down anything that comes to mind, even if it seems small and silly or too grand and unrealistic.

Also, consider other areas you can explore (education, VR world, business, hospitality, etc.) to gain new ideas. Looking into other fields doesn't mean copying what the secular world does, but you can gain valuable insight from some of their strategies and practices. There's a reason the most successful businesses have been successful, and leaders can learn from some of their processes and approaches. Just think how much more effective church leaders can be when they combine successful strategies with the gospel – their ministries will be unstoppable!

If nothing was holding you back (not even a lack of budget or volunteers), what would you want to see in your ministry this year regarding digital discipleship? We serve a God who specializes in doing the impossible, so dream big!

Take time now to start your dreams list and keep it handy as you read the rest of the book for more specific ideas.

Action Steps

1. Write down your ministry's vision statement.

2. Draw out a ministry funnel or discipleship pathway and add the events, programs, initiatives, etc. to each stage. This can be for your entire ministry or focused specifically on digital discipleship.

3. Complete a SWOT analysis of your ministry when it comes to digital discipleship.

4. Start your brainstorming, blue sky dreams list regarding digital discipleship, then keep it handy as you continue reading this book.

Don't forget to check out the resources page online at deeperkidmin.com/timetoupdateresources for tools and printables to help with your action steps!

Chapter 2

Safety and Integrity in Digital Discipleship

THINK OF THE GUARDRAILS on the side of the road on highways and especially in mountainous areas. You probably pay them little attention or barely notice them as you take in the views and keep your eyes on the road. They're uneventful and don't add to the scenery.

Ideally, you will never experience the necessity of a guardrail, so they continue unnoticed as you drive to your destination. But guardrails provide safety and security when things go awry.

Policies and procedures act as the guardrails for your ministry. They're not flashy, can be tedious, and tend to blend into the background, but they provide a proactive opportunity to protect all ministry members.

Digital Ministry Policies and Procedures

Safety and integrity are vital to a healthy children's ministry, and they apply to your digital ministry strategy, too. So, how can you keep your kids, families, and volunteers safe while engaging online?

First: Put the guardrails up and create a digital ministry policy for your ministry. You probably have procedures for keeping your kids, volunteers, and staff safe in other areas, and your online presence shouldn't be any different. A digital policy defines and outlines the purpose of digital ministry while also setting guidelines and boundaries to ensure the safety of children, volunteers, and staff.

1 Peter 5:2-3 encourages leaders to *"be shepherds of [their] flock"* with integrity and care, which would include doing everything they can to ensure the safety of those they lead. Paul's letters to the church in the New Testament show intentional thought in church-building policies and procedures. These same proactive approaches apply to your digital ministry as much as your in-person ministry.

Policies and procedures build trust and protect your people. Pastors and churches are not exempt from legal responsibility for their actions, and well-defined protocols help protect not only the children but also the staff and volunteers. Leaders must make sure their staff, volunteers, and elders receive support in case of a lawsuit.

Check with your senior pastor and other staff to see if the church has a digital ministry policy. That way, you can ensure your children's ministry policy complements the church policy. If your church doesn't have one, start the conversation about creating one.

When crafting your digital ministry policy, think about identifying the following:

- What can be shared online (on the website, social media, etc.)? Are there any pictures, topics, or information you want to avoid sharing on social media or the website?

- Who is allowed to share it? Who has access to the admin side of the church website? Can volunteers post pictures from Sunday mornings on their personal accounts? Who will help you manage the ministry's social media accounts?

- When is it allowed to be shared? For significant announcements, should social media posts be scheduled before, after, or at the same time an email goes out?

- How should the varying social media platforms be monitored and by whom?

- Should or how should your staff and volunteers use the internet and social media to interact with minors or parents online?

Especially think about including specific digital ministry guidelines for your volunteers and have these expectations as part of your volunteer training.

Volunteer Policy Ideas:

- Ask volunteers not to post about kids on their personal social media accounts and instead share the pictures with you as the children's ministry leader. On one occasion, one of my preschool volunteers posted a picture of one of our more rambunctious preschool boys running by in a blur on her Instagram account with a caption that said things were "crazy but so fun" in the preschool that morning. The volunteer, of course, didn't intend anything hurtful, but when the mom of that preschool boy saw the post, she was so embarrassed. Situations and struggles at home heightened the mom's sensitivity to the post, and she felt humiliated and ashamed. If a potential visitor saw that post, they may feel uncomfortable sending their more reserved preschooler to a class that seems "crazy" and super high-energy.

- Volunteers also may not know every child's background and family history, which can lead to legal and safety issues for kids, especially those in foster care or unique family situations. I served as a volunteer at our church's annual Easter Egg Hunt one year. I posted pictures of kids running across the field to hunt for colorful eggs on the Deeper KidMin Instagram account without thinking anything of it. A few hours after the event, I got a call from the children's pastor, kindly explaining that some of my pictures included children in foster care going through difficult court situations whose pictures could not be on social media for safety reasons. My children's pastor was incredibly gracious with her explanation and requested that I remove the photos, which I did immediately. I was enjoying the event and trying to share what a great job my children's pastor had done in planning it, but I had inadvertently put a child at risk through a simple post.

- Of course, you can only police some things your volunteers post online. But if you have them send the pictures to you or a designated team member rather than them posting on their social media accounts, you have more control over the images used and the messages conveyed.

- Unless their specific volunteer role requires or includes it, ask all volunteers to stay off their phones, social media, and other digital platforms while serving with your ministry. That helps them focus more on the kids and lets you monitor what and when things get posted. It also allows them to be more present and engaged. Consider asking youth to leave their phones in a safe, designated place while serving.

- Give volunteers guidelines for how to interact with kids and parents online. Just like you wouldn't want a volunteer alone with a child behind closed doors in person, ask that all digital communication with a child under the age of 13 be done in a public forum or with another adult involved, and the initiation of the communication should always come from the younger party, not the volunteer. Some churches even choose to discourage any connection online at all. Evaluate what level of digital boundaries works best for your ministry and communicate those boundaries to your volunteers.

- Remind your volunteers that their online presence reflects them and your ministry. For this reason, it is imperative that they conduct their online interactions according to your volunteer handbook or code of conduct.

Then, before finalizing your digital ministry policy, ask an attorney to review it. Legal assistance helps ensure you comply with all state laws, especially those relating to mandatory reporting, which can vary by state.

After you've crafted your digital ministry policy, be sure to share and communicate this policy with parents, volunteers, and staff members. Many guardrails have reflectors or additional signage on or around them that help notify drivers of their existence, even in dark or difficult driving conditions. Policies are only effective when you share them with everyone involved. I recommend making this part of your annual and ongoing volunteer training. Sharing policies helps develop a unified vision and consistent messaging.

Sharing policies also allows you to enforce them better. When you discuss the digital standards at your volunteer training, consider asking volunteers to sign a digital or social media agreement as part of their volunteer agreement. The document gives your ministry the right to dismiss a volunteer who doesn't adhere to the policy.

Finally, review your digital policy periodically. Highway guardrails need inspecting every three years to ensure they are well-kept, still required, and secure.[11] I recommend checking all your policies and procedures, including those for digital ministry, every year, but at least every three years.

Best Practices

Once you establish your official digital policy, consider a few best practices for safety and integrity in the digital world.

Have permission to post pictures of your kids and volunteers. I always include a media clause in our annual registration and new kid information forms. You can find the verbiage of that media clause on the Time to Update resources page online at deeperkidmin.com/timetoupdateresources. Families can opt out of this part of the release. By updating our medical release form to a *media and* medical release form, we ensured we had written and signed permission to use pictures of kids online. Most families were OK with it, and it helped our team be more aware of any children who could not be in pictures online, especially children in foster care or unique family situations.

For volunteers, you can include a media clause within your volunteer handbook or volunteer covenant. Asking permission from your families and volunteers to share pictures of them online shows your respect for them and your diligence in doing your part to keep them safe in the digital world.

Don't share personal information about kids or families in your posts. Avoid using kids' names, age-specific descriptions, grades, sports teams, neighborhoods, schools, and other identifying locations. Even a simple post with a caption that says, "I so enjoyed watching Haley in the 3rd-grade talent show today at Jackson Elementary!" shares too much personal, potentially harmful information about that child.

If kids are wearing name tags in your pictures from a Sunday morning or another church event, a good rule of thumb is to blur or black out the text on their name tags to help protect them. It takes some extra work, but it helps you ensure their safety.

Only use content that you have permission to use. You should have legal, copyright permission to post all pictures, graphics, images, videos, etc., on your website or social media. I'm not saying you need written permission for every piece of media you share, but be aware of where your content is coming from, and give credit if you're using someone else's work.

I recently found out about a church using the graphics from one of my Easter events without permission (meaning they didn't purchase the resource) and without giving Deeper KidMin credit (they tried to blur out the Deeper KidMin logo). Using someone else's pictures is stealing and it feels very personal to content and graphic creators. You wouldn't walk into an art store, take a picture of your favorite painting, print the image, then frame it for your ministry or claim it as your artwork.

The same mentality also applies to digital artwork (and other digital resources). By complying with the licensing and terms of use of any resources (Deeper KidMin or other), you help protect digital intellectual property and support organizations and creators to continue providing the children's ministry community with affordable and creative resources.

For royalty-free pictures and graphics, I like using Pixabay, Pexels, or Unsplash, and Canva is my go-to place for creating graphics. As of 2023, Canva even offers a free subscription to their "Pro" level account for nonprofits.

This approach of integrity applies to the videos and audio you use, too. YouTube clips can be so helpful, but many children's ministry leaders aren't aware that playing videos from YouTube is a breach of YouTube's terms of use, which state: "You may not use the Service to view or listen to Content other than for personal, non-commercial use (for example, you may not publicly screen videos or stream music from the Service)."[12]

The solution? Paul Clifford from Twelve Thirty Media encourages churches to get permission from the creator before using a video from

YouTube. "Find out who owns the video, which might admittedly be difficult, and get their specific, written permission to show the video in church, in the way you want to," he says.[13] Leaders must also be careful about the movies, movie clips, and even music they play, ensuring they have the licensing and permission to play them in a public setting.

A helpful resource that provides more detailed solutions and licensing for using media with integrity is Christian Copyright Licensing International (CCLI). CCLI provides guidance, licensing, and more for churches to use, play, and stream various forms of media while helping ease the burden of the administrative aspects of copyright laws and usage.

The world of copyrights and licensing may seem frustrating and overwhelming, and that's because it is. Unfortunately, there isn't a one-size-fits-all approach or a simple three-step process to ensure you're using digital content correctly, especially since the policies vary with each platform, tool, and curriculum.

But the Church must set the example of sharing digital content with integrity by ensuring leaders have permission to share or show digital content and by giving credit to the original creator when applicable.[14] Speak with other staff members about what they're doing regarding media licensing. Look at your internal practices and systems to check whether you engage in the digital world with integrity.

Ultimately, when in doubt about a piece or source of digital content, don't use it. If you need clarification on whether a piece of digital content is OK to use or show, stay safe and find an alternative.

Action Steps

1. Start drafting your digital ministry policy. Use the template provided in the Time to Update Resources page online at deeperkidmin.com/timetoupdateresources to help you get started. Or, if you already have a social media policy in place, review it for any updates.

2. Make a plan to educate or remind your staff, volunteers, or families about the policy, whether through a special training, social media posts, group texts, or a simple reminder email. This policy will provide the framework and guardrails to keep you on track as you integrate digital discipleship into your children's ministry strategy.

3. Have conversations with your staff and do research to ensure you have the necessary licensing or complete the required reporting to use digital content with integrity.

Chapter 3

Your First First Impression–Your Website

No, that chapter title wasn't a typo. There's a new first impression that visitors have of your church, and they see it before they even walk through your doors: your website. Research has shown that 75%-90% of visitors will visit your church website before attending.[15]

When my husband and I were in a season of transition and looking for a new church, the first thing we looked at was the church's website. And honestly? We quickly judged whether the church would be a good fit for us based on the website. The churches whose websites looked like they were made in the 90s and hadn't been updated since then went on our "not for us" list.

In Lead Digital, an online course all about digital ministry, Josh Denhart shares:

> A close friend recently moved to a new community. They are committed Christians. She said, "We are going to this church for the first time. It is time to see if who they are online is really who they are as a church." Stop everything and listen to what she said and what it means for us as leaders. The "get to know you period" for newcomers has changed. The pathway from non-attender to a core member of your church has been transformed and elongated. This is the age of the internet. People will do their research long before darkening the door of your church.[16]

Your website offers a significant opportunity to connect with first-time guests and committed families, and it should be an intentional part of your digital discipleship strategy. Does your church website have at least one page dedicated to the children's ministry? If you have a children's ministry page, what kind of first impression does it give to potential first-time guests?

I know that web development and control over the website may be out of your hands as a children's ministry leader, but you can (and should) still have a voice in making sure your *first* first impression is a good one.

Four Questions to Help Evaluate Your Website

1) Is Your Website Up to Date?

If your website looks old, has low-quality photos, or includes outdated content, people will assume your church is old-fashioned, low-quality, and obsolete. You might think high-quality websites require a high-quality budget, but an effective website is possible, regardless of size or financial limits. Let's narrow the focus down to your children's ministry page in particular:

Are the pictures of kids who are still in your ministry? Or, are they pictures of kids who are now in high school? Use real-life images (not stock photos) of your ministry whenever possible, and update these pictures at least every two to three years.

Are the dates and information updated, including ministry times, locations, and events? People use your website to gather information about service times, children's ministry offerings, and more. If it's March and you still have information about the children's Christmas production as your focus, then it's time to update those details.

Have you checked out what other church websites have to offer on their children's ministry pages? I'm not suggesting you copy what other churches do, but look at some well-known churches in your area or that you are familiar with and spend some time exploring their website and the children's ministry section. What are some elements of these websites that you like? Did you see something you want to avoid? Seeing what other churches are doing with their website can help you determine what to do with yours.

2) Is Your Website First-Time-Guest Friendly?

On average, a web user will give your site 10 seconds to engage them before they click away. When looking at a church's website, most people want information, like service times, location, what to expect, childcare, etc., so make sure your website quickly and clearly communicates these details. Many churches are now making entire sections purely for visitors and first-time guests to access information quickly.

Consider the questions a first-time family might have about attending the children's ministry at your church. Do you answer those questions clearly and quickly for them in the top section of your children's ministry page? Include details like classroom locations, ages, ministry mission statement, and details about checking in your first time. You could include a digital copy of your first-time guests' form so visitors can complete it online before they come on Sunday. See an example of a first-time-guest digital form on the Time to Update Resources page online at deeperkidmin.com/timetoupdateresources.

Sliders (an automatically switching carousel of images that rotates through a series of images) are very trendy on websites, but they can be confusing for first-time guests. Imagine that a guest is reading about your Sunday morning children's ministry program when suddenly the image changes, and now they're reading about the Christmas program coming up in two months. They're not interested in the Christmas program, and now they need to learn how to return to the information they were reading before. Then the slide changes again to something about the "At-home Discipleship" Parent Bible Study you're offering. The first-time guest is just trying to learn where and when to show up on a Sunday morning, and they're not ready to dive into a parenting Bible study with you.

You may also want to consider the language you use on your website and assess whether someone who doesn't attend your church (or isn't a believer) can understand it. For example, using themed names for your children's ministry ages or classrooms is a fun and creative way to set up your children's ministry environment. But a first-time guest doesn't know the difference between the seeds and sprouts class. So use general, universal language, or explain the language that is unique to your ministry.

Ensure that your website is set up for the families who already attend your church and for the potential families who you *want* to attend. Being user-friendly is an essential element of a digital discipleship strategy.

3) Is Your Website Easy to Use?

When was the last time you spent significant time with a website that was slow to load and confusing to use? The brain naturally chooses the easiest path, so cumbersome and confusing websites encourage people to click away and go elsewhere.

First, think about the navigation of your website and your children's ministry webpage. Can users find your children's ministry page quickly and easily from the site's home page? A good rule of thumb: A user should be able to find the information they need about your children's ministry in three clicks or less. If finding your ministry information takes more than three clicks, your website needs to be simplified, because you're most likely either frustrating or losing people.

Your children's ministry page should also encourage a specific action from visitors. Calls to Action (CTAs) are directional messages like "Learn More," "Sign Up," or "Get Involved." They help website visitors know where to go or what to do next. An easy CTA to start with is the first-time-guest form mentioned previously. If you're promoting an upcoming parent and child small group on your children's ministry page, an inviting "Learn More" button encourages visitors to click the button to find more information about the small group. Or your goal might be to get visitors to register for an upcoming event, so you'd use a CTA like "Register" or "Sign Up." Or if you send out weekly or monthly email newsletters, a simple "Stay Connected" button invites people to join your email list. CTAs guide church members and guests to complete the actions you want them to make, ultimately allowing you to better connect with them and create touchpoints that lead to future interactions.

Next, think about the layout of your children's ministry page. Once they arrive at your children's ministry page, is the information presented clearly and easily understood? Studies show that when the human eye looks at a website, it scans the content in a "Z" pattern: left to right, top right to bottom left, and left to right again. Line up the most important information on your children's ministry page within the "Z" to minimize information overload.

Use large, easy-to-read fonts to grab attention and give focus to your most important message. Remember, you will want to use a nice, clean web-friendly typeface. Even though a funky typeface might feel more kid-friendly, it takes more work to read, especially for people browsing your page from a mobile device. Most people surf the web from their phones, so if your site doesn't adapt gracefully to a mobile-friendly version, it might be time to give it an update. Using large images, large headings, and simple fonts can help make sure your ease of use is still high, even on a mobile device. So, when evaluating your children's ministry page, explore the page from a computer and a phone to ensure maximum ease of use.

4) Does Your Website Communicate Your Vision?

Parents want to know that there is a safe, purposeful, and fun place for their children. Your children's ministry webpage should give a glimpse of what it's like to be part of your ministry. Share the vision not only in writing but in pictures, also. Use pictures of your ministry classroom and environment, not just when the rooms are empty but when they vibrate with the energy of kids worshiping, playing, and learning together. Show pictures of kids smiling, having fun, and interacting with their friends and leaders.

In addition to sharing the important information and visual glimpse of your ministry, share the wins and celebrations. People like to connect with people, so consider a quote or story from a parent, volunteer, or a child in your ministry about how that ministry made a difference in their life. I even saw one church create a video where they asked kids, "What's your favorite thing about (ministry name)"? This video lived at the bottom of the children's ministry page on their website to help share their vision and make connections with new families.

Your vision should also include resourcing and equipping parents, too. Your children's ministry webpage is a great place to offer a digital parent resource library with suggested links, books, articles, ideas, and more to support your parents in the discipleship of their kids.

Sharing the purpose and vision of your ministry visually and through testimonials on your website helps guests get a feel for your ministry and encourages them to take the next step. When you connect with families online, the potential to connect with them in person grows.

Evaluate Your Site

Once you use these questions to evaluate your children's ministry webpage as a staff or on your own, test your site. Ask a younger and an older couple in your congregation to find specific information on your site and survey them with questions like: Was the site easy to navigate? Was the information you needed easy to find? Was the description of the children's ministry true and accurate?

You may think the website is out of your control and not your concern, but everything about your church's website reflects on your ministry.[17] Work with the site developer on your staff to update your children's ministry page every couple of years and ensure it is easy to use, welcoming to guests, and a solid representation of your ministry. You only get one chance to make a great first impression. As leaders work to reach a generation of digital natives, it's their responsibility to ensure that the *first* first impression–their website–is excellent.

Action Steps

1. Look at your website with fresh eyes. Can you answer "yes" to the four questions from this chapter? What conversations need to happen to update your website, if needed?
    - Is your website updated?
    - Is your website first-time-guest-friendly?
    - Is your website easy to use?
    - Does your website communicate your vision?
2. Recruit three or four diverse people to explore your website from an outsider's perspective.

Don't forget to check out the resources page online at deeperkidmin.com/timetoupdateresources for tools and printables to help with your action steps!

Chapter 4

Teaching Digital Natives

ONE DAY WHILE HELPING me clean out my home office, my toddler found one of my old laptop computers. She immediately slid it in front of her on the floor, opened it, and began tapping on the keys. I haven't touched it in years, and I don't think it even turns on anymore, so I let her play with it. She spent almost 45 minutes "working on projects" and even having online meetings about planning for VBS. (Can you tell she's a ministry kid?) She even grabbed her Minnie Mouse cell phone while working to "text a friend" and take a selfie with her new computer.

I never sat her down and explicitly taught her how to use a laptop or a cell phone. But she knew exactly how to play with that old computer, because she's seen me working on mine. She's seen me use my phone to text, play music, and take hundreds (OK, thousands) of pictures of her and her sister. For my daughters and every kid in your ministry, this technologically connected, digital world is all they know. Technology and internet access aren't special to them; they're just part of life. Like a desk and chair are regular fixtures in offices, technology is the same for today's kids.

The children (and even parents) in your children's ministry are digital natives, meaning they do not know life without the Internet. Marc Prensky, the creator of the term "digital native" defines them as "native speakers of the digital language of computers, video games, and the Internet."[18] Technically, digital natives include anyone born after the 1980s, though some argue that the first actual generation of digital natives was born after 1995.

Anyone born before the 1980s is known as a digital immigrant, meaning they learn about technology as it comes out at an older age rather than digital media being part of their reality since birth. Like immigrants of all kinds, digital immigrants adapt to their environment but still retain their "accents" of the past. Some examples of digital immigrant accents may include printing a document to edit it, calling to ask if you received their email, or even making a phone call vs. sending an email in the first place. These accents influence the way digital immigrants operate and the way they teach in the children's ministry world.

Digital natives have spent their entire lives surrounded by computers, video games, cell phones, streaming services, and all the other tools and apps found in the digital world. They learn through observation how to use these technology elements as they watch their grown-ups use them. Their interactions with the digital world are changing the very structures of their brains and influencing the way they learn. Because of the ever-changing, constantly connected digital world, kids will have very different childhoods. They will enter adulthood with very different experiences from adults of previous generations.

The former ways of teaching and children's ministry are not bad or wrong, but they taught a different set of learners. As a teacher of the gospel, you must recognize the differences in the way digital natives learn, relate, and process information, and you must update the way you teach to better engage digital natives with the Bible.

Teaching Kids, Not Lessons

So when teaching digital natives, what would you say is the most crucial aspect of your weekly lesson? The "hook"? The gamification elements (more on this later)? The life application?

Plot twist! One of the most critical aspects of your lesson doesn't even have to do with the lesson itself. It's the *kids* you teach and how well you know them! Effective and creative Bible teachers place as much value on understanding the children they teach as on the lesson's content. Understanding kids now includes knowing how digital natives react to learning, interaction, and relationships.

Solid content and preparation are essential, but if you're not taking the unique characteristics of the digital natives you teach into account, your effectiveness drastically decreases. This need for knowing your audience is a widely acknowledged concept in public speaking and teaching. As the digital world continues influencing how kids learn and grow, it's time to update your approach in ministry, too.

Shift your mindset from teaching a Bible lesson to teaching the specific group of children God has placed in your ministry. That means you must know the kids you teach intellectually and relationally, especially as their digital native status influences each aspect of who they are and how they learn.

Intellectually

Intellectually, the very wiring and physical makeup of a digital native's brain is different than in generations past, and they need very different opportunities from a learning environment.[19] Marc Prensky encourages teachers to think of digital native kids like rockets.[20]

Rockets are fast. First, digital natives need speed. Not only do they learn and process information at high speeds, but even the very schedule of their lives is faster than ever before. Digital natives are born multi-taskers, often switching between tasks and ideas quickly. Consider providing fidget toys and physically interactive elements during the teaching time to engage these multi-taskers while also helping them learn how to focus. Ministry leaders must wrestle with this tension between keeping up with the fast pace of the world and assisting kids in slowing down to experience and respond to the Holy Spirit.

Digital natives are also able to find the answers to their questions quickly. "Let's look it up!" is a phrase my toddler says often when she asks a question I don't know the answer to. Right now, her questions usually center around the name of a movie character or the eating habits of a bear, but one day, she'll have more profound faith questions. I love her spirit of curiosity to take charge and find the answers, but I also recognize that where she asks the question can significantly impact the quality and truth of the response.

The voices offering answers online are not always faith-based; ministry leaders must teach kids discernment amid their high-speed chase to discover the answers to their questions. By connecting kids with peers and other trusted adults within the church, you can encourage kids to ask questions in a safe space and share other trustworthy resources to explore at home.

Rockets explore. Second, rockets are exploratory by nature. For digital natives, learning is triggered by internal curiosity and spurred on by boundless imagination. Sometimes in children's ministry, leaders get so focused on the biblical knowledge they want kids to learn that they forfeit the development of kids' biblical imagination. You may think knowledge and imagination are mutually exclusive, but they're really two sides of the same coin. Imagination spurs a curiosity for knowledge. Knowledge allows for expansion of the imagination.

Imagination is central to learning and intellectual development, especially for digital natives. Experts who research and explore this method of imaginative education define the imagination like this: "It is the ability to think of the possible, not just the actual; it is the source of invention, novelty, and flexibility in human thinking; it is not distinct from rationality but is rather a capacity that greatly enriches rational thinking."[21]

Too often, leaders treat the Bible as something to memorize and defend, not as a great mystery that learners are called to discover and participate in. When you invite a sense of wonder into your teaching, you harness a powerful learning tool that fuels creativity and innovation for the digital natives you teach, giving them opportunities to explore the Bible, not just learn about it. Curiosity invites questions. Curiosity stirs wonder. It helps kids connect their heads and hearts by tying in knowledge with emotions.

Curiosity and the chance to explore on their own help kids see familiar stories in a new light that inspires an internalization of and transformation from the lesson. Effective Bible teachers won't just teach kids the stories of the Bible or strip down the lesson in the name of simplicity and life application, though life application is important. Leaders must help kids explore the extraordinary within the stories of the Bible so they can discover redemption, wonder, and mystery.

So, what can leaders do to incite curiosity and encourage biblical imagination when teaching digital natives? Share the story (and The Story) – *"The Word of God is alive and active,"* and leaders should teach its stories well (Hebrews 4:12). Leaders need to invite kids into the action of the plot to join the people of the Bible on their way to the story's climax. Think about some of your favorite storytellers and utilize some of their techniques when teaching. Also, try to show the smaller Bible story in the context of the overarching Story of the Bible. Where does it fit in? Go beyond just "Old Testament" and "New Testament" or even "before Jesus" or "after Jesus" and talk about how the story fits into God's redemptive plan. Help kids make connections between the stories to see how they connect as God's Big Story, chronologically throughout history. I know some leaders avoid using the word "story" to help differentiate the Bible from fairy tales and fiction, but Hebrews also tells us that Jesus is the author of our faith. He was a storyteller Himself. Framing the gospel as the best story ever told invites kids into the action.

Laugh as you learn. The easiest way to reduce boredom and encourage engagement is to make the lesson fun and enjoyable! Digital native kids are used to "edutainment"–a combination of entertainment and education–where learning is intentionally fun. Include jokes or funny observations about the story. There are lots of things in the Bible that make me laugh. For example, Matthew 4:1-11 tells us that after 40 days of fasting in the desert, "Jesus was hungry." Ha! I'd be hungry, too, after 40 days of fasting! This line always makes me laugh, but when I dig deeper into that laughter and seemingly obvious detail, I see that this detail shows that while Jesus was fully God, He was also fully human. He hungered, He thirsted, and He felt pain. The silly "duh" moment leads to a deeper understanding.

Another great way to help digital natives learn is to tap into their feelings. Feelings and emotions lead to remembering and learning. Not only does emotion affect attention (kids will pay more attention the stronger they feel about a story), but emotions can enhance long-term memory retention. The simplest way to connect feelings is to share: a) how you think the people in the Bible story might be feeling (either based on the text or context clues), or b) invite kids to share how they would feel if they were a person in the Bible story.

Identify and acknowledge the unknown. Present the story as part of a great mystery or adventure. There are many things we don't know about the Bible, and that's OK. Draw attention to the unknown or unexplained parts of the story. It's OK to leave some things unanswered. Unanswered questions help kids learn that faith is a lifelong journey and allows them to explore on their own.

Be filled with wonder yourself. As the teacher, you can't take kids where you have never been yourself. So, do your imaginative deep dive into the Bible story as part of your preparation for teaching the lesson.

Rather than jumping straight into the script the curriculum gave you, read the Bible story yourself. Underline or highlight parts of the story that evoke emotions, make you chuckle, or cause you to wonder about something. If *you* are swept into the imagination of a story, that will shine through as you teach kids, inviting them to use their creativity and imagination, too.

God can do *"immeasurably more than all we ask or imagine,"* and He's given us minds to think creatively and wonder curiously at His Story (Ephesians 3:20). The minds of digital natives are even more curious and creative than ever because of the endless opportunities they have to learn from and connect with the world around them. When leaders intentionally invite wonder and curiosity into the learning process with digital natives, they help kids cultivate a biblical imagination that shapes a more robust biblical knowledge and leads to life transformation.

In line with their exploratory nature (and the developmentally inquisitive nature of elementary schoolers), digital native kids would much rather puzzle out the answer on their own than be told the correct answer. Rather than telling them the main point of the story, present them with questions and help them explore the answers in the Bible. Consider reversing the review time and the teaching time. Ask the questions up front, then use the Bible lesson or story to help kids discover the answer. For example, rather than telling kids the three promises God made to Abraham in a large group setting, invite them to explore the Bible with their small group, identify the three promises, then move back to the large group to discuss their findings.

Think of digital natives as researchers, thinkers, and explorers, with you as the guide, goal-setter, questioner, and coach that helps point the rocket in the right direction. Ask questions that allow for exploration and discussion, not just questions with one specific correct answer. While some content review questions require simple, concrete responses, go beyond factual questions and into open-ended questions that allow for personal and practical replies. Ask "why" questions or

"I wonder" questions. This type of questioning will enable kids to develop their biblical imagination as well as their biblical knowledge.

Tap into digital natives' curiosity by allowing kids the freedom to explore and discover the lesson independently. Let's shift from the grown-up expert teacher telling kids what to think to exploring the story together, valuing what kids notice and share. Modifying teaching this way may require adjusting your small-group ministry and large-group teaching time. Both contexts are still highly valuable. But both need some modification to engage digital native learners and allow them to explore. Remember, when their curiosity is piqued, their engagement will be too.

Like rockets that need to launch into space before truly becoming rockets, digital natives learn best by doing and experiencing, not by listening, so create time and space for kids to be involved in the lesson. How can you invite kids into the teaching of the lesson itself? Invite kids to act out the story, solve a puzzle, consider a question, stand up and complete an action, vote on what they think will happen next, provide sound effects, or various other interactive storytelling techniques.

Remember those "choose your own adventure books" that gave the reader some say in how the story ended? Apply that mentality to your Bible teaching. Offer "simulations" of the Bible story and give kids options of what they think will happen next. Then, explore the results of their choices and teach what really happened. Think less lecture and more partnering when it comes to teaching digital natives. Shift from a grown-up talking at kids to multi-sensory and experiential learning.

In the spirit of doing, engage their bodies and not just their minds. Use gestures, play charades, or invite kids to make faces as to how they would react if they were a person in the story. What pose might represent the different people, cities, places, and actions happening in the story? Seek patterns, look for repetitions, or put it to music: "What would it be like if this part of the story sounded like a drumbeat? If this part of the story was a sound, what would it be?" Speaking of sounds, engage the senses as much as you can. How would different parts of the story feel, touch, taste, smell, or sound? For example: ask kids what the ark might have smelled like.

And if you can turn learning into a game? Even better! All kids love games, but this generation of digital natives finds games the preferred way to learn. They enjoy pursuing and acquiring knowledge when it's a challenge to overcome rather than a list of facts to memorize. There's no reason kids can learn 100 Pokémon cards but not learn the books, stories, and people of the Bible. It all depends on the presentation.

According to research, using games in teaching can help increase a child's participation and foster social and emotional learning. All kids need this after two years of a worldwide pandemic that left us isolated and socially distanced.[22] Kids can engage with the lesson (rather than merely listen) when gamification is involved, and games can help keep their attention and focus. Game-based learning gives kids ownership over their spiritual growth and development, improving enjoyment and overall retention of what they learn.

When I say "gamification," I don't mean adding another game to your children's ministry schedule. I mean incorporating game-like elements into the teaching, like goals, obstacles, and learner-chosen cause-and-effect options. Can you give kids a goal or specific task to accomplish at the beginning of the lesson that requires them to participate, engage, and explore during the lesson to meet that goal? What markers, checkpoints, or points can you provide to let kids know how they're doing on their quest to accomplish that goal? Will they need to work together or "play" individually? Whether you invite the whole room to participate in the game elements or have just a few kids come up on stage, gamifying your teaching time will help you connect with the digital natives you teach.

Use role-playing games during your teaching time to encourage kids to use their biblical imaginations and explore different perspectives. Turn the Bible lesson or review time into a competition or gameshow setting between small groups or halves of the room. Set up a simple points and rewards system for kids who complete specific tasks. Many leaders already do something like this with Bible Bucks or a Bible Points store.

Jordan, a kidmin leader in Texas, took this idea one step further and is exploring ways to host her ministry's Bible Bucks store online, where kids can earn and redeem their points in a digital space that makes it easier for Jordan and her team to track it.[23]

Beyond the in-church experience, point kids and families to Bible games online and other Bible apps they can use at home.

So, whether the lesson itself becomes a game or you gamify the review time, finding ways to incorporate play into the learning process will help you better connect with the digital natives you teach and will help them retain more.

Gamification in your ministry *can* go too far and become toxic or counterproductive. You don't want kids to think that their faith is only a rewards-based system where God will grant X if they accomplish Y. If you forfeit biblical depth in the name of gamification, it's gone too far.

Full disclosure: I'm still exploring and experimenting with this, too! If you have any ideas that have worked for you, I'd love to hear them!

Rockets require letting go of control. Third, leaders must recognize that they can aim rockets in the right direction, but they don't have complete control over their course or speed once they launch. Digital natives hold the same great potential for discovery and exploration, but leaders must let go of some control. As a control freak myself, this is scary! But it's so valuable to engage digital natives with the gospel. How can you give leadership and responsibility to some of your older kids to help you teach the lesson? What would it look like to invite kids to create the teaching elements of the class or divide the story among your small groups and let the small groups devise a way to teach the rest of the group?

One of my favorite ways to release control is through prayer stations or a worship response time. During a worship response time, there may be one to three prayer activities available for kids to participate in, and they get to choose which one(s) they want to do. The activities vary in type (and sometimes noise level), but they all connect to the teaching in some way or provide kids the opportunity to process the lesson with God. They enable children to express their responses to God about what they are learning and who He is.

Each station looks different and allows the child to respond to God differently. After the lesson, kids are released for 5-10 minutes to complete the prayer station(s) of their choice independently, at their own pace. The first time I tried this, I thought complete chaos would ensue and we'd end up with a room full of crazed kids. But they surprised me, and the Holy Spirit worked through this time to speak to kids' hearts in new and profound ways.

Some of my favorite prayer stations that work for any lesson include:

- Listen quietly to or sing along with a worship song.

- Use an old cell phone to talk to God about what you're learning.

- Draw a picture of your prayer or something from the lesson.

- Write a prayer on a sticky note and stick it to a cross.

- Read through the Bible story on your own and draw a picture of who you'd want to be in the Bible story.

- Walk through a simple labyrinth (made from a bedsheet or tape on the floor) as you talk to God about something from the lesson.

- Use Scrabble tiles to make a board of words that relate to God or the lesson.

- Shape pipe cleaners into the initials of someone you want to pray for or into people from the story and reenact the story.

- Write a note to God, asking Him your questions or telling Him what you love about Him.

I always communicate with kids about the importance of this time when doing prayer stations. I let them know up front that worship response is a time for them and God, and I ask them not to distract their friends. Volunteers help monitor the room by redirecting kids who are getting off-track and by helping non-readers. Find more examples and resources for prayer stations on the Time to Update resources page online at deeperkidmin.com/timetoupdateresources.

Is letting go of control scary? Yes. But is it worth seeing kids connect with the Lord through interactive learning and prayer? Absolutely.

Ultimately, digital native kids have huge potential. Just as a rocket's mission to space can provide astronomical results, when you embrace the rocket ship analogy of how digital natives learn, God can do immeasurably more than you ask or imagine with the next generation.

Relationally

Digital natives learn best when they feel known, seen, loved, and heard by whoever is teaching them. They must understand that their opinions matter. Kids want to feel valued for their ideas, or they will check out. Generation Alpha, born from 2010 - 2024, want to know if you know them and pay attention to their likes and dislikes. The same will be true for Generation Beta, the upcoming children who will be born from 2025 - 2039.

Build relationships with your kids, both inside and outside of Sunday morning. Building rapport increases your effectiveness as a Bible teacher. Be OK with times of fellowship and fun in your ministry that allow for connections and relationships to form.

Learn kids' names as quickly as possible. Know what your kids are into and what they enjoy doing. I don't care about Popits or unicorns, but the second-grade girls in my small group are obsessed, so I prioritize knowing about them too. At the beginning of each school year, consider asking kids to fill out an "All About Me" page that invites them to share their favorite candy, books, movies, games, restaurants, etc., then keep these sheets on file, share them with your volunteers, and review them from time to time to remind you about the kids you serve. (You can find a free printable About Me page on the Time to Update Resources page online at deeperkidmin.com/timetoupdateresources.)

If your church has access to devices for small groups, share the About Me questions in a digital survey as a small group activity using Google Forms, Survey Monkey, or Poll Everywhere. Kids love answering questions and then immediately seeing the survey results from their peers.

Then, take their favorites and incorporate them into your teaching. Do you notice that several kids mention playing soccer? You could use a soccer ball as part of an object lesson or as a teaching tool. Do lots of kids say they like Skittles? Then use Skittles as a reward for a game or to help you teach the lesson. Are there children who enjoy drawing? Make drawing the Bible story part of your next teaching time.

Digital natives are used to customizing and personalizing everything, and when leaders can personalize the teaching time to match kids' interests, it shows how much leaders value them. When you know what makes kids excited in their everyday lives, you can tap into those passions when you teach to better engage them with the gospel.

Digital natives also want to know how their learning connects to the real world and how it applies to them in real life. Always incorporate child-friendly examples of life application into your teaching time and invite other real-life people into the teaching. This may look like volunteers who share their faith stories, other leaders within the church, guest speakers, or a FaceTime call with a missionary.

When kids see people living out their faith in their daily lives, they explore ways they can live out their faith in the context of school, home, sports teams, and more. Today's children are much more receptive to teaching when it is applicable and obviously relevant.

As you're thinking about life applications, consider how the Bible also applies to the digital world. For instance, is the lesson on words? Then mention the importance of the words they type and post online. Encourage kids to wonder why God gave us the ability to communicate in such innovative, powerful ways through words and technology. Talk about the different ways we can use technology to tell the story of Jesus Christ. Ministry is all about the people, not programs or curricula. Incorporating kid-specific passions into teaching time is the perfect opportunity to demonstrate how the Bible is relevant and applicable to kids today.

For digital natives, many interactions with friends, family, and teachers occur online or through a screen. They are used to connecting with people in various ways and have access to communication nearly 24/7. The digital world also allows them to communicate with people beyond their physical location, opening a broader network for learning and growth. Digital natives are constantly connected with others, and this influences their preferred learning styles.

When learning, digital natives prefer groups and collaboration to lecture-style teaching. How can you adapt your teaching time to include opportunities for small groups to explore the Bible story together as they learn it, rather than just as a review time after you've taught the Bible story from the platform?

Look at the ways you are setting up the space. Where and when can you provide spaces for kids to interact more often rather than sitting in rows and listening to a lecture? Is there room for kids to move around, explore, and discuss? Consider sitting in a large circle or having small groups sit in circles. Albert Einstein said, "I never teach my pupils. I only attempt to provide the conditions in which they can learn." Be intentional about the conditions you provide for kids to learn about Jesus.

Think of yourself less as a curator and provider of biblical knowledge and more as a leader or facilitator of learning experiences. Creating group-centered learning and opportunities for collaboration (with leaders and with their peers) helps create an ideal environment for digital natives to learn. One easy way to invite collaboration is to encourage kids to ask questions in your ministry and throughout the teaching time.

We practice full immersion baptism at our church, so we dunk you. You go all the way in and come all the way out. We bring the children into the service for baptisms so they can see their friends and other church members get baptized.

A few years ago, we were having a baptism day, and my children's pastor asked me to hang close to one of our first-grade girls. She was a foster child with a very tough story, and had difficulty focusing, sitting still, and engaging. She was also a runner. (You know the kind.) So my children's pastor asked me to sit with her during the baptisms.

That Sunday was her first time seeing a baptism, and she was very concerned about the people who were getting baptized. She had lots of questions:

"Do they have to put their head underwater too?"

"Do I have to get baptized?"

"Wait, why aren't they wearing a bathing suit?"

"But Ms. Brittany, won't their clothes get wet?"

Me: *(Whispering, trying not to distract those around us in the front of the sanctuary but also trying to honor her curiosity about what she was watching.)* "Yes, sweet girl, their clothes will get wet."

Child: "EVEN THEIR UNDERWEAR?"

She blurted out this question in such horror and with such volume that all the other kids around us heard, and at least half the sanctuary. I tried to help bring her volume down while continuing to answer the non-stop questions she had throughout the baptisms. "Why do you get baptized?" "Will they be wet forever?" "Is the water cold?" "How do they get that big tub in the sanctuary?" "How old do you have to be to be baptized?" She was more engaged in those 5-10 minutes of being able to ask questions about what she was seeing than I had ever seen her.

As concrete thinkers trying to discover the world and their place in it, digital native elementary kids have *many* questions. But they don't often feel confident enough to say them out loud. When kids know they can ask questions without consequence or judgment, it builds trust, and it's healthy for kids to see that sometimes adults don't have all the answers either.

You and your team can encourage kids' questions with a simple routine that lets them know it's OK to have questions about the Bible, faith, and church.

Before beginning your regular review questions or other small group activities (or even partway through teaching the lesson), ask this simple question: "Do you have any questions about today's Bible story?" Remind kids they may want to ask questions if they are curious about a part of the Bible story that made them wonder about something or if they were unsure what it means.

Give kids a few moments to think and ask. If no one has questions, continue with your regular activities or the rest of the lesson. If someone has a question, take the time to listen and respond. Invite other kids to share their thoughts or responses to the question. If you know the answer, share it with the group. If you don't know the answer, say so! It's OK to say, "I don't know! But I'll get back to you next week." When kids, especially digital natives, see that adults don't have all the answers either, they learn that faith is an ongoing journey. You must offer space for digital natives to explore the questions kids have about their faith so you can help kids discover the answers. Questioning allows kids to process and integrate their faith with their daily lives.

P.S. That same little girl who had all those questions about baptism ("EVEN THEIR UNDERWEAR?") got baptized just last year. Watching her questions and concerns turn into understanding and belief was an honor and a sweet privilege.

Collaboration may also look like inviting kids into the creation of the lesson. Digital natives love to create, and I'm not just talking about crafts here. Kids want to use their skills and abilities to create something with intentionality and purpose as part of their learning process, not just as a way to review it. Pause during teaching time and provide craft supplies or blocks. Let kids develop their own craft to represent the story or use the blocks to build what's happening. You could even ask kids to create the props for the story while you're telling it.

Invite kids to create the graphics (or even simple drawings) for the teaching time or give them time to make a video with their small group that retells the Bible story. Training kids to help run the technology or soundboard allows them to uniquely serve and engage with the lesson. And let's be honest, most of them can figure out how to adjust the sound better than many adults can anyway.

Collaboration and relationships are vital to the learning process for digital learners. So, when you find ways to help them connect with others and the content, you'll more effectively engage their brains for learning.

Using Technology in Teaching

Once you understand digital natives, how they learn, and the importance of connecting with them relationally, there are other simple ways to incorporate technology into your teaching.

Examine your curriculum. See what digital resources (videos, slides, graphics) the curriculum provides, and then take a closer look at some of the large-group and small-group activities. Where can you add a digital element to the large or small group time?

Our preschoolers always enjoy using the Bible App for Kids on a ministry tablet to interact with a Bible story after hearing about it in the main lesson. Elementary schoolers love to make a video of them reciting the memory verse or listening to a song of the memory verse instead of just writing it down. Consider allowing your preteens to use the Bible app on their phone as you study Scripture together, and especially check out the Kids Bible Experience within the Bible app for a kid/preteen-specific daily devotion tool. You may not have the budget for each small group to have their own iPad or tv screen in their small group space. Instead, you can offer a station rotation model that allows groups to rotate from space to space rather than staying in one classroom. Rotating through stations allows groups to use the main screen for small group activities.

Use your screen. Use graphics on the screen or even sound effects in the background during your teaching times to help you teach the Bible story and lesson. Digital natives prefer graphics over text. Even if your graphics are just a simple picture of the Bible story, having a visual on the screen will help digital natives focus. And every time the graphic changes, it helps kids re-engage in what's happening up front.

You can also use your screen to play a video. While I don't recommend letting a Bible story video teach the lesson for you, it's a great supplement! Maybe you teach the class in person, then use the Bible story video during snack or small group time. That way, kids hear the story twice and in two different formats or ways. Also, consider utilizing lyric videos for worship time, countdown videos to get the session started, or video-led games during dismissal.

Jessica Bealer, an author and well-known children's ministry leader, said in an interview with me about digital discipleship: "Videos are here to stay. I can't stress this enough: children's ministry leaders need to be using video as a tool in their ministries."[24]

Jessica makes a vital distinction with this statement: videos and other technology are excellent *tools* you now find in your ministry toolbox. Digital media provides opportunities for you to better connect with digital natives and support this new teaching shift from lecture-based to partnership teaching. You can use these technology tools to supplement and complement your teaching times, not as the primary source.

And don't forget the families who join you for Sunday services from home. Ask yourself if sharing your Sunday mornings in children's ministry in an online format is a possibility or need for your ministry. You may need more resources or tools to livestream your children's ministry service or activities, so consider other ways to connect with families online. Consider recording the lesson ahead of time and airing it "live" on Sunday morning via a social media platform or your church's website. Offer a Zoom meet-up for kids after church on a Sunday afternoon, allowing you to connect with kids and families who don't physically attend your church.

Here are a few tips for engaging kids on video, whether pre-recorded, live-streaming, or even in a Zoom meeting:

1. Make a Plan - Just like you walk into a regular Sunday morning with a plan for the day, do the intentional planning for a video, too! Make sure you and your team know the video's order and flow, and set any props off-screen but within reach. You may also want to have links ready to share, like links to activity worksheets on Google drive, links to worship videos, links to a video lesson, and more! (Make sure your curriculum & worship songs allow sharing the resources in this way.) Spend the time prepping for this video just like you would for a typical Sunday, and remember to communicate that plan to your families and volunteers.

2. Have a Teammate - Ask another team member or a spouse to help you host the video (especially if you're going live). Whether this teammate is on-screen with you or handles things behind the scenes, an extra person helps you focus on the kids rather than the technical side. This person can help with any technical difficulties, pass you props as needed, be your tag-team partner to maintain engagement on-screen, monitor comments as they come in, and post any necessary links in the comments for families.

3. Check Your Tech – Whether you're using fancy equipment your church owns or setting up your laptop at the kitchen table, keep these things in mind:

    - Framing – Set the camera at eye level when possible; this

may mean stacking some books underneath your computer or raising the tripod a few inches. Make sure you are centered in the middle of the frame and have enough space around you for anything you have planned (games, walking back and forth, or crazy arms for the Bible lesson). Your object lesson won't be as effective if half of it is missing from the screen. Part of making your plans should be to decide if you'll be sitting or standing. This will make it possible for you to frame the video appropriately.

- Another tip: Do your best to look at the camera lens, not the screen (especially if you're using a laptop)! It's challenging and feels weird, but to the viewers, it makes it seem like you're looking directly at them.

- Lighting – You want to be brighter than your background. So, if you're sitting in front of a window, make sure you're looking out the window rather than having the window behind you. You may also need to move a desk lamp closer to your set-up or turn off lights behind you to help prevent you from looking like a dark silhouette. Consider investing in a ring light or another light source made specifically for video.

- Sound – Do your best to minimize background noise in your space, and if you're in a big room, pay attention to any echoing that may make it hard for kids and families to hear you. The further away you are from the camera, the higher the chances you'll need a microphone.

4. Check Your Delivery – Have you ever watched a kids' show on TV or YouTube? The people on those shows all tend to overemphasize their tone of voice and inflections, with dramatic pauses and facial expressions. Learn from them! Be just

as energetic as you would in front of a room full of kids. Will it feel awkward doing this in front of a camera? Yes. But push through the awkwardness (you've got this!), and though you will feel weird, kids will see the zany children's pastor they know and love!

5. Make it Familiar and Interactive - Follow your regular "schedule" as much as possible. Kids thrive on routine, and the familiarity of your church time will help them feel connected even if they aren't with you in person. So begin with your usual activities, say the phrases they hear you say each week, and make it fun and interactive! Add games that encourage kids to complete actions (like Simon Says, just without anyone getting "out") or games that invite kids to answer questions in the comments. Ask questions throughout the video that encourage kids to respond by sharing their answers with someone they're watching the video with or by submitting responses in the comments. Do things to act out the Bible story or invite kids to make "sound effects" for the Bible story. Find ways to encourage interaction, even through digital platforms.

One children's ministry leader even started including "listening stations" in her preschool classroom, where a child could access a set of headphones, a Bible storybook, and an audio version of the Bible or the day's specific story. Preschoolers could then hit play and "read" the Bible themselves through the audio version.

Using technology to reach digital natives is more than just popping in a Bible lesson video and calling it a Sunday. It's using the technological and digital world we live in to engage kids in the greatest story they'll ever learn.

Technology has impacted how kids learn and interact, providing a new tool for a more robust teaching and learning experience. It gives digital natives more ownership of their faith and spiritual formation.

Technology often allows kids to interact more with the story and engages more of their senses in learning. When you tap into kids' natural inclination toward the digital world and incorporate some specific intellectual and relational nuances about this digital generation as you teach, then you'll be better able to connect them with the gospel in profound and life-changing ways.

Digital Curriculum

"And please don't let the kids touch the cellos."

That was a sentence I once had to include in our teaching curriculum as a reminder for our volunteers. We were a mobile church that met at a local high school, and the children's ministry happened in the band, orchestra, and chorus rooms. Every piece of the curriculum we shared with our volunteer team had to be modified in some way to fit our context.

Currently, I'm a children's ministry volunteer at a church with its own building, but my children's pastor still has to modify the lessons based on materials, service times, and the needs of the kids and families we serve. No matter what curriculum you use, you're probably modifying it in some way to meet your families' needs, too. As technology advances and provides new ways for leaders to communicate and share information, there are more and more options for creating lessons that allow you to share the gospel in every ministry context.

One tool that continues to grow in prevalence is digital curriculum. While digital curriculum at its most basic level is familiar, the ministry world continues to expand its offerings in digital curriculum.

Because of the ever-changing nature of technology, the term "digital curriculum" doesn't offer a clear definition encompassing all types of curriculum offers. Instead, digital curriculum tends to fall into a progression of formats.

The earliest forms of digital curriculum provided leaders with a way to purchase and access files through a provider's website with the option to download the files to the leader's computer. This level of digital curriculum often allows leaders to share the files digitally with their team through the provider's website, sometimes even creating accounts for their volunteers to access the curriculum online, too.

Other curriculum providers expanded the definition of digital curriculum to include video-led teachings and presentation elements within the teaching files, allowing leaders to use the files on-screen while teaching.

Recently, curriculum providers have started exploring a new form of digital curriculum where leaders can access the lessons and media within the provider's website, application, or portal. Rather than merely downloading the document and media files, the lessons can be manipulated, edited, and formatted within the leader's account on the provider's site. This option provides a seamless way for leaders to modify the lessons for their context while maintaining the overall professional formatting (without taking up too much space on their hard drive).

This last option of digital curriculum offers leaders the ultimate control in preparing lessons for their specific context. A leader can adjust the individual activities included, easily rearrange the order of activities, modify the overall time length, and even type their own wording into the professionally formatted document. This new digital curriculum gives leaders all the main pieces of a puzzle. Rather than the pieces only fitting together in one way to form one specific picture, the leader can maneuver and adjust the puzzle pieces to create a lesson plan most compatible with their preferences.

After leaders have edited the content to be exactly what they and their kids need, the options for sharing include:

- Downloading the files to their computer.
- Emailing them to their team.
- Adding team members to the portal itself.

- Sharing directly to a screen with a presentation view that allows leaders to see their notes while projecting the media on-screen for kids.

Choosing a curriculum that offers this new level of digital accessibility allows leaders to adjust lessons as needed without losing the professional look of the formatting and without requiring lots of time to make the changes. As with any new technology, there is a learning curve in interacting with a new digital curriculum interface, but leaders who take the time to learn will save time in the long run.

Regardless of the digital curriculum level, leaders can reap the benefits of digital curriculum. Digital curriculum in all forms allows for easier sharing with your volunteer team. You can share the lesson files with them digitally through email, scheduling systems, and even private social media groups. This digital access allows volunteers the chance to preview the lessons beforehand and sometimes even teach the lessons from a digital format.

Digital curriculum can help leaders save physical space and resources. With digital curriculum, you print only what you need and prevent extra booklets from stacking up in your office or storage closet.

The digital curriculum platforms now available for leaders have come a long way regarding user friendliness and ease of use. Most now include video tutorials and guides on how to use the platform most effectively, and their seamless processes allow for easy exploration and implementation of the materials. Gone are the days of crossing out content on printed materials or re-typing content to modify it. Digital curriculum is becoming easier and more accessible for leaders to use.

Ultimately, digital curriculum allows leaders to create lessons and provide worship experiences that best fit their ministry context. I've never met a leader who used a curriculum or lesson plan exactly how it was provided. We all modify the activities or wording to best reach the kids in our ministry. And that's a good thing! God has placed the families you lead in your care for a reason, and you know them best.

Digital curriculum provides leaders the opportunity to modify the content to better serve the kids and families in their context.

As you explore options and the feasibility of using digital curriculum in your ministry, consider your:

- Budget – Can you afford the digital curriculum options? How do they compare to your current budget for curriculum, including printing and material costs?

- Time – How time-consuming is your lesson prep each week? In what ways could using a digital curriculum decrease your time spent on modifying lessons?

- Team – Are your volunteers ready and willing to explore digital options? Do you have a few key leaders willing to "test drive" digital curriculum options so they can share feedback and suggestions?

As the world at large streamlines its processes and systems for sharing information, the processes and communication systems in ministry will also continue to adapt. I foresee more and more curriculum providers offering digital curriculums exclusively over printed materials. You can embrace the tool of digital curriculum to help you better equip and empower your teams to teach kids about Jesus in the way that works best with the flock God has called you to shepherd.

Update Your Teaching Strategies

Whether you're teaching kids in a large-group setting on Sunday morning, a small-group mid-week setting, or anything in between, make an effort to learn about this generation's learning styles and the specific needs of your group. Take advantage of the tools technology offers to update your teaching strategies. When you partner with the rocket ships in your ministry to help them explore the gospel, you create learning experiences that develop faith that lasts. When you shift your teaching times to a partnership experience, kids are more engaged and transformed by the gospel, leading to lifelong discipleship.

Action Steps

1. Evaluate your next teaching time by asking these questions:
    - How can I be more mindful of how digital natives learn as I prepare to teach?
    - In what ways am I currently partnering with kids in the learning process, and how can I do that more effectively?
    - What part of this lesson allows me or my leaders to connect with the kids? What part of this lesson will enable kids to connect with each other?
2. Brainstorm three methods or ways to make your teaching time interactive over the next month.
3. Identify the digital resources provided with your curriculum and evaluate if you are using the digital resources provided to their fullest potential.

Don't forget to check out the resources page online at deeperkidmin.com/timetoupdateresources for tools and printables to help with your action steps!

Chapter 5

Partnering with Parents

IF YOU'VE BEEN SERVING in kidmin for a while, you know that children's ministry isn't just about the children. It's about their parents and grown-ups, too. And just as leaders must recognize and adapt their teaching styles and ministry approach with digital native kids, they must also acknowledge and adjust how they connect with their parents.

The first generation of digital natives, the millennials, are now the parents in your ministry. Many of your flock as a children's ministry leader, from kids to their grown-ups, are now digital natives. As digital natives grow up, have children, and bring them to your ministry, they will also bring their technological habits and preferences to the table. This reality should influence how you communicate, encourage, and equip the parents and caretakers in your ministry.

Communicating with Parents

Digital native parents most likely prefer digital methods of communication, like emails, text messages, app notifications, or social media. In the education world, research has shown that weekly text messages about a child's academic performance with tips for at-home practice "led to increased communication with teachers, better performance in school, and fewer absences."[25]

I'm not saying you should start texting out weekly spiritual report cards for the kids in your ministry, but what would it look like to connect with parents via a mass texting app like Remind or Text in Church? You can schedule text messages in advance and connect with parents throughout the week. I have used Remind to text parents some discussion and review questions each Sunday right after church, in addition to sending home take-home sheets every Sunday. That way, they could easily see what the kids learned and have a simple way to engage in a faith conversation over lunch.

If you still want to send home take-home sheets, maybe you create a small postcard that reminds parents to check their text messages, the ministry social media account, or that has a QR code that links to your website where parents can find the day's video lesson, at-home ideas, and more.

QR codes can also be beneficial for new families who attend your church. Rather than giving them a full pamphlet with pages of details about your ministry, provide a simple postcard with a QR code that links to your ministry website page, a particular landing page for new families, or even a LinkTree page with links to multiple access points and information.

In a Facebook post in the *Kidology.org – Let's Talk KidMin* Facebook group, Karl Bastian, creator of Kidology.org, shared a simple, digital idea to follow up with visiting families: text replacement and texting. [26] He set up text replacement on his phone, meaning when he types in a specific phrase or set of letters, this message automatically populates in his texts:

> "Hello NAME! This is NOT an automated text, this is actually me, Pastor Karl, the Kids Pastor from Highpoint Church on my phone reaching out to say WELCOME – delighted your family joined us for church. I hope KIDNAME had a super time! To keep this text short I created a page with some info & short videos to give you a glimpse into our kid's ministry. If you have a quick minute, click here: Highpointchurch.us/kids/sneakpeek If you have any questions, don't hesitate to reply!"

Then, he replaces "NAME" and "KIDNAME" with the appropriate names of the visiting family members. "It's been very successful. I get lots of friendly replies," Karl said about the strategy. Since texting is the most frequently used form of communication among Americans younger than 50 according to a Gallup Survey from 2014, this idea of texting new families demands consideration.[27] And by setting up the text replacement shortcut, it takes less than two minutes to send a text like Karl's example and follow-up personally with a visiting family.

Though texting shouldn't be overlooked, email is still a standard mode of communication. As of November 2021, approximately 95% of internet users in the U.S. report using emails as a primary method of online communication.[28] With the capability to schedule emails or send emails to a particular group of people (like all your preteen parents), it is a valuable tool for communication with parents about anything and everything in your ministry. Consider sending a weekly family email outlining upcoming events, special reminders, and the upcoming lesson topic and big idea. Choosing to share the following Sunday's lesson details instead of the previous Sunday's allows parents to begin faith conversations at home and position themselves as the primary disciplers of their kids.

With the ability to automate or schedule emails ahead of time, you can prepare a month's worth of weekly emails or even use automated emails for special events. For any major event, and especially one like VBS, I scheduled emails to go out the day before and each afternoon of the week, with reminders for parents about that day's lesson, the mission offering item for the next day, and a preview of the next day's lesson. I don't know about you, but I don't have much energy once VBS is over each day (there's no tired like VBS tired), so I scheduled these emails ahead of time. That way, parents still got the valuable information they needed, but I didn't have to use precious energy in the afternoons after VBS to concoct the email well. Finding ways to utilize email in both on-demand ways and to schedule them ahead of time can be a great way to maximize your busy time as a leader and communicate with parents through the platform they prefer.

You may not have everyone in your community's email, but you can use digital means to communicate about outreach events. We used a Facebook event to share about a drive-through Christmas Nativity event and had over 1,500 people RSVP online that they wanted to come. On the night of the event, about 750 drove through, and people from our church and the community were commenting, posting, and sharing about the event on their social media profiles. It was a huge success for our church. Facebook events make it easy for people to share about the event and invite their friends. Facebook events also make it easy share important information with anyone interested in attending.

It's also time to rethink how you do sign-ups for events and collect payments from parents. Gone are the days of paper sign-up sheets and cash placed in envelopes. Digital native parents prefer to sign up for events online and, most importantly, pay online. Paying online may require a specific setup and often includes transaction fees, but the ease of use for parents to pay online is invaluable. Today's parents rarely carry cash, and only some know where their checkbook is. Instead, they rely on online payment methods for most of their transactions, and many set up recurring payments (even for tithing and charitable gifts) to automate the process.

Providing a way for parents to sign up for events and pay online makes it easier for them and truthfully, more manageable for you as the children's ministry leader. With digital sign-ups and payment methods, there's an instant record of who has signed up, who has paid, and other details. Possible solutions for creating online sign-ups include Planning Center Registrations, Google Forms, Sign Up Genius, and Jot Form, to name a few. These systems provide the capability to collect payment and information, then quickly organize it into spreadsheets that help you track the results.

After events, consider sharing pictures of the events through digital sharing methods like Shutterfly, Flickr, or Google Photos. Parents love seeing pictures of their children. And if you can share photos from a special event or even a Sunday morning that parents can easily save, share, and post themselves, that's an easy way to include them even if they weren't there.

My daughter's preschool class has its own photo sharing site, where the teachers upload pictures from school each week. I get a glimpse into my daughter's day at school, and when there are cute ones, I can save them and share the pictures with others. I love having access to these little moments in my daughter's day that I would otherwise know nothing about, and it creates a natural conversation starter at home. I can show her a picture from the week and ask her what she's doing. It instantly creates memories for her and discussions for us.

You can also utilize the digital world for following up with families after events, including asking for feedback or sending out surveys. Ask families to complete a quick survey or even respond to a few simple questions in an email. Using JotForms, Google Forms, and SurveyMonkey allow you to easily create digital forms and collect the results to help you improve the event next time.

Using digital methods to communicate with the digital native grown-ups you serve may take some adjustments of processes and habits. But, when you share in the way they know best, you are more likely to get a response and foster communication in return.

Encouraging Parents

The ability to communicate with parents and speak into their lives creates the opportunity for children's ministry leaders to encourage parents, too. An easy way to keep up with the families in your ministry is through social media platforms like Twitter, Instagram, YouTube, and Facebook. Now more than ever, leaders can find out about what's going on in someone's life without actually having to talk to them. And while the implications of this change are vast, one positive is that you can quickly receive updates on the lives of the families in your ministry. Use social media to see their family outings, celebrations, grievances, and even remember family birthdays.

Adopt the attitude of Romans 12:15 in your social media interactions with families. Romans 12:15 says, *"Rejoice with those who rejoice. Mourn with those who mourn."* When you see someone celebrating something on social media, leave a comment to let them know you're celebrating with them! If you see someone mourning or going through a tough time, leave a comment to let them know you're thinking of them and praying for them, and consider following up privately to minister to them in their time of need.

Leaders can pray digitally for their families, too. I heard about one church that hosted weekly mom's prayer groups over Zoom, and these short, 30-minute meetings provided a space for worn-out moms to share and pray over each other during nap times or in between loads of laundry. Hopefully, you are already intentional about praying for the parents in your ministry, but using digital methods of asking for prayer requests and sharing prayers reminds parents they are not alone in the discipleship of their kids and in their own journeys of faith. Text them and tell them you're praying for them. Post on social media and invite them to share a prayer request with you via a private message. Tell parents you are thinking of them and praying for them as often as you say the same thing to their children.

As parents navigate their world, countless voices speak into their thoughts, lives, and parenting styles. While the stream of social media images can cause comparison, discontent, feelings of failure, and more in a parent's mind when they compare their day-to-day to someone else's highlight reel, ministry leaders have a unique opportunity to infuse some scriptural encouragement into parents' social media newsfeeds. Be a voice that helps take the pressure off. That reminds them of their identity in Christ. You can be a light that contradicts the condemning and comparing voices of the world to remind them of their role as parents in an encouraging way.

You can also use social media and the digital world to connect parents to each other. Create communities online and remind parents and caretakers they are not alone. Helping parents connect with others in the same stage of life and a different stage of life helps remedy the lack of competence many parents feel. Whether it's small groups, preschool playdates, family events, Facebook groups, or something else, incorporate ways for parents to build relationships with other parents.

Facebook groups can be an excellent tool for helping parents connect. Nona Jones' book, *From Social Media to Social Ministry,* is an excellent resource for starting your Facebook groups. Consider creating groups for parents of littles, parents of preteens, grandparents in your ministry, or volunteers.

I spoke with Dede, a ministry leader in Georgia, about how she used the ideas in Jones' book to utilize Facebook groups in her ministry .[29] She tested some of the ideas with a target audience of first-time grandparents, which led to a thriving ministry even in the middle of the pandemic. They offered an event for first-time grandparents and, through typical church communications, got only three people to register. She waited a week, then started a closed Facebook group for first-time grandparents, which grew to just under 100 members within the first 24 hours. She promoted the same event within that group for grandparents and got 15 new registers within 36 hours and an invite to bring the program or event to two other churches in her area.

Encouraging digital native parents may include motivating them to step away from the digital world now and then. What would it look like for everyone in the family to go screen-free for one hour a day and one day each week? Andy Crouch, author of *The Tech-Wise Family*, says putting technology away allows families to connect better and be more present with each other.[30] Kids and youth are learning their tech habits from the adults in their lives, and when adults set tech boundaries, it helps kids learn to develop tech boundaries, too.

Whether parents realize the full extent of this reality or not, their interactions with the digital world shape their worldviews and self-identities. Children's ministry leaders can use the digital world to speak life and encouragement into an often lonely space.

Equipping Parents

Encouraging parents easily leads to equipping them in their roles and faith. You don't have to become an expert in knowledge if you are an expert in resources. Be familiar with the digital resources available for parents so that when they ask, you can give them a short answer that still provides all the necessary information. Give thought to a parent resources page on your website with suggested books for various topics or life stages and a link to your Facebook groups where parents can talk more and share life experiences. You can even utilize the features within a private Facebook group to share files or articles or create a LinkTree page for your Instagram bio with links to helpful parent resources.

Another source of knowledge for parents could include resources more specific to your ministry, such as a monthly calendar of Sunday's Bible stories that parents can read at home with their children before coming to church. Weekly text messages or social media posts with discussion questions for mealtimes or short examples of how to use various daily routines (bath time, brushing teeth, etc.) for a faith conversation may also be helpful resources. By providing ways that allow kids and families to interact with Bible stories from Sunday mornings throughout the week, ministry leaders essentially enable families to *"tie [the stories] as symbols on [their] hands and bind them on [their] foreheads"* (Deuteronomy 6:8).

Use social media and other digital methods to offer simple ideas for faith conversations at home. Many parents feel overwhelmed by the thought of at-home discipleship, let alone the pressures to parent a certain way, feed their children certain foods, and enroll their children in certain activities. When you step into the digital world where parents spend much of their free time, you can provide easy ways for them to develop faith habits within their families.

The key to equipping today's parents is simplicity, but simple doesn't have to mean shallow. Simplicity may look like sharing one Bible story video with families, then one or two activities for them to complete at home that use minimal supplies and require little (if any) printing. In a viral Tweet from 2020, Scott Irlbacher, a higher education administrator in Pennsylvania stated: "Quit sending millennials things to print at home... We don't have a printer. [We've] been surviving on secretly using our work printers for years." This Tweet had over 1,000 replies and over 200,000 likes, showing the number of people resonating with the Tweet's message.

Providing ideas to parents that require extensive time, thought, and supplies can often have the opposite effect of what you hope for, because parents feel too overwhelmed. I talked with Jessica Bealer, a leading voice in children's ministry, about equipping parents digitally, and she said, "Parents don't want to feel like they're letting their child(ren) down.[31]" So, when parents feel encouraged with simple and easy-to-use resources, they are much more likely to implement them.

In my early years of ministry, I provided resources about how to be the best parent and what the "Christian parent" should be doing in their home. I quickly learned that sometimes these resources created more guilt and shame than encouragement or advice. Moms felt they could only live up to some of the resources I provided, leaving them feeling overwhelmed and inadequate.

I learned to switch my mentality as a children's pastor from "parent coach" to "parent cheerleader." It may seem like semantics, but this shift creates a brand-new mindset to interacting with parents and completely changes your approach.

So, what's the difference? A coach is someone who instructs or trains and gives advice on improving. A cheerleader is an enthusiastic and vocal supporter of what someone is doing.

Rather than being another voice in the world that tells parents what they should and shouldn't be, find out what parents are doing and come alongside to support them. Consider what parents are already doing at home and brainstorm ways you can help them bring faith conversations into those moments and activities.

Nightly family prayer at bedtime? Consider sharing prayer prompts for families to use to guide this prayer time. Are families heading out to lunch after church? Text them with some discussion questions to talk about while they eat. Are parents pulling out after-school snacks? Share ideas for connecting with their kids in these pivotal moments after school. Are parents driving in the car to soccer practice? Provide a playlist or share an audio devotional that families can listen to in the car.

My church's children's pastor created a Spotify playlist of the ministry's favorite worship songs so we can listen to it in the car, and it's my go-to playlist for any drive with my daughters. As families spend more and more time in the car, look for creative ways to equip them for those in-between moments of travel in their day.

For some of our special events, like VBS and church-wide weekend retreats, I created podcasts for families to listen to that talk about the day's lesson or provide conversation starters. If you do not create podcasts of your own, are there any podcasts you'd recommend to the parents in your ministry?

Deanna, a children's pastor in Georgia, included a "Biblical Parenting Moment" in her weekly newsletter emails.[32] This short section provides encouragement and practical ideas that equip parents for faith discipleship at home. Equipping parents to add small faith habits to their family's daily routines can help ease the pressure many parents feel. When they see how simple it can be to incorporate faith conversations into their daily lives, they'll be more likely to have them.

Another easy way to equip parents is to play to the rhythm of the family's life. The families in my ministry shared that May and December are some of the busiest and most stressful months of the year. So, I intentionally didn't plan anything in May, and I made sure the only thing I added to their December calendar was a Christmas-themed Parents' Night Out that gave them a little break. Whatever the busy months are for your families, make sure you know them and can serve them digitally during those seasons.

Be aware of back-to-school and testing seasons, school breaks, and other important holidays. Consider ramping up your digital offerings during these seasons as families are less likely to attend church or may be more on the go than usual.

As digital native kids have more and more access to information at the swipe of a finger, consider equipping parents with resources that address lifestyle and everyday topics from a biblical worldview. Kids are exploring sexual identity, life with technology, and their sense of self at younger and younger ages, leaving most parents feeling lost as to how to navigate these difficult conversations. How can you equip parents to feel confident discussing these issues with their children? Whether you offer seminars or classes (online or in person) or provide digital resources through a parent page on your website, you can equip parents for faith conversations, even the challenging ones.

Brandy Thixton, a children's ministry leader in Tennessee, shared with me about the Parent Collective events her ministry hosts two or three times a year. These Parent Collective events are an online, live-stream event geared toward resourcing parents. Her most recent one focused on mental health with kids and teens and featured speaking from Dr. Andy Yarborough, a licensed clinical psychologist and founder of The Well Clinic. These Parent Collectives are available on-demand after the event, and the church selects topics based on the parents' questions and concerns. Brandy shared that one surprising "win" from these Parent Collectives was the reach and engagement, with over 1,200 views. "There's no way we would have gotten 1,200 parents in the church that night, and through this online setting, we actually get to reach parents that otherwise wouldn't or couldn't get to our location," Brandy said.[33]

As parents try to navigate the digital world for themselves and their kids, consider offering guidance or training about technology with kids and faith milestones regarding media. This might look like a family technology agreement, apps and habits to help families stay accountable online, or helping parents navigate conversations about online boundaries. (There's a free downloadable sample of a family technology agreement on the Time to Update Resources page online at deeperkidmin.com/timetoupdateresources.)

In 2021, I attended an online webinar from a local church about online safety for kids, even as young as the toddler years. I listened from the comfort of my home office (and, let's be honest, in my pjs), and the webinar provided excellent resources for me to continue exploring as my daughters grow. Ask the grown-ups you serve what topics their kids are bringing up at home and how you can better equip them to navigate those conversations.

With all the resources you provide parents, remind them constantly that your role is to come alongside what they are doing at home, not to burden them or add something else to their plate. You're already working to connect with the grown-ups in your ministry to partner with them in the discipleship of their children. So, as you consider your digital discipleship strategy, explore ways to serve adults digitally, too.

Action Steps

1. Do you know how your parents prefer to communicate (email, text, take-home sheet, etc.)? Take time to ask them, and then make a plan to implement or improve one form of communication through digital means.

2. Send a text to three different parents in your ministry, letting them know you're thinking about them and praying for them today.

3. Research one new resource you can share with families to equip them for discipleship at home. Need a place to start? Check out the list of faith-based apps on the Time to Update Resources page online at deeperkidmin.com/timetoupdateresources.

Chapter 6

Connecting with Volunteers

Ah, volunteers. Volunteers are the most common pain point when I talk with children's ministry leaders, mainly because no one ever seems to have enough of them. Even when there are ample volunteers, leading people who lead other people comes with its own unique set of challenges.

There are entire books written about volunteer strategy in children's ministry, but let's hone in and focus on utilizing technology with volunteers. Whether you use texts, emails, vlogs, blog posts, or social media, utilize the digital world to connect with volunteers.

A robust volunteer strategy will include intentional plans, procedures, and systems for recruiting, coordinating, training, and investing in volunteers. And fortunately, the digital world can make these purposeful plans and systems more effortless than ever.

Recruit Volunteers

Don't you wish a million hands would shoot excitedly into the air when asked if anyone wants to volunteer in the children's ministry? Me too! Unfortunately, that's not quite how it works, so strategically think through recruiting volunteers and inviting them to join your kidmin team.

Start with prayer. As with everything else in ministry, start with prayer. Acknowledge God's ownership of the ministry and ask Him to *"send out workers into His harvest field"* (Matthew 9:37-38). Begin by praying about the ministry as it is currently, where you feel God is calling the ministry to go, and who can help get there.

Identify the needs. Prepare a list of positions needed within the ministry and create job descriptions for each area. You must know what you're looking for to make the recruitment process effective and successful. Look ahead to the future of the children's ministry and determine any current or new positions you will need to reach ministry goals. Does the digital world present any new needs within the ministry? Would a videographer, photographer, or social media host be helpful? What new roles might someone take on to help in implementing a digital discipleship strategy? Ensure volunteers and delegation are part of the digital ministry strategic plan. Like other areas of ministry, leaders must tackle the world of digital discipleship with others.

Share the vision. Communicate and actively share the robust vision of the children's and family ministry rather than merely talking about the people needed to help. Recruit from an individual viewpoint by asking how a person might use their gifts to support the ministry instead of simply striving to fill roles. This person-centered approach will help people catch the purpose and the passion behind children's ministry, encouraging them to join the team. You could even go deeper and use a spiritual gifts test to help each member within the congregation find their strengths, then place them in positions that meet their passions and abilities. Effective leaders will adopt a kingdom focus when recruiting volunteers, seeing the bigger picture rather than just holes that need filling. Especially if you're launching a new digital strategy for ministry, communicating this vision and how it will help your ministry and kids grow is essential.

Build relationships. Announcements from the stage and inserts in the bulletin asking for volunteers are a good place to start, but effective volunteer recruitment requires much more than an up-front call. Leaders must get out from behind their desks and talk to people in the congregation. Learn about their unique gifts, passions, and strengths. Do some "background research" on social media profiles to see what interests someone and how they spend their time. Recruiting steadfast volunteers demands recruitment procedures like that of Jesus. He built disciples first, then asked them to step up and lead.

Be excited. When you communicate the need for volunteers in your children's ministry, especially in the digital world, show your enthusiasm and excitement for serving. The way you talk about serving in ministry will directly influence how people think about serving in ministry. Use your gifts of inspiration to share the vision of kids' ministry and to help people see their essential role on the digital team. Leaders inspire those around them by communicating their vision enthusiastically and capturing followers' hearts. Admit when you need help or training from someone who knows more about the digital world than you. When volunteers have ownership over their roles and see their purpose, they'll be as excited as you.

Wait patiently. Sometimes your volunteer roster takes time to fill with the right people. Be patient with the recruitment process, and feel free to wait for the right person to fill the right position at the right time.

Keep going. Even when you feel all the needs are met, keep sharing the vision of the children's ministry and inviting people to join the team. Recruiting volunteers is a never-ending task and should always be part of your ministry strategy, whether digital or not. You never know who God is going to place on your team!

In her Mastering Ministry Workshop Facebook Group, Jessica Bealer shares about two churches that utilize QR codes in their volunteer recruitment process. The first church, Canoe Creek Christian Church in St. Cloud, Florida, made small business cards with a simple question, "How can I serve?" alongside a call to action to scan the QR code. The QR code takes you to a form to add your name, info, and an indicator that you'd like to serve.

What would it look like if you passed out those simple business cards once a quarter, once a month, or had them readily available every week? Staff members can quickly and easily follow up with anyone who submits the form, helping them get involved right away.

Willow Creek Community Church in Chicago, Illinois used QR codes to help share details about specific volunteer roles. They displayed their volunteer opportunities in the lobby area, and each position had a page that cast the vision for the role and a QR code that led to more details, including a complete job description and how to get plugged into that role.

QR codes are simple and free to make, so the only cost associated with this idea is the cost of printing. Check the Time to Update Resources page online at deeperkidmin.com/timetoupdateresources for a free Canva template to create a personalized QR code recruitment flyer.

Speaking of recruitment, you may need to recruit new volunteer positions to help with the digital discipleship aspect of ministry. Consider creating new volunteer roles like photographer, content creator, social media coordinator, and digital hospitality team.

- Photographer - Does someone in your church, including a teen, take great pictures? Invite them to attend one Sunday morning a month in your ministry to capture the morning in pictures and video. You don't need a fancy camera for this volunteer; most smartphone cameras provide high-quality photos. And I'd recommend adding "event photographer" to the list of volunteers needed at every special event. Having consistently reliable pictures allows you to communicate the vision of the ministry and invite volunteers to be part of a child's faith formation.

- Social Media Coordinator - Invite someone to run and manage the ministry's social media accounts. Be sure to meet with them beforehand and address parameters, the graphics standards guide (example and template found on the Time to Update Resources page online at deeperkidmin.com/timetoupdateresources), and any other specific boundaries. Then release the social media management to them and take it off your plate.

- Content Creator - Even if someone only takes over part of the social media coordination, you can ask for help creating the content you will post and other digital content you may need. Share the graphics standards guide with a few volunteers who enjoy graphic design. Then, let them create the graphics and videos for social media and teaching times. There are probably kids, preteens, and youth in your church who enjoy digital content creation. What a great way to teach them how to use their skills to serve!

- Digital Hospitality Team - Members of your digital hospitality team intentionally interact on social media with ministry posts, ask and answer questions to foster conversations, share posts for exposure, and monitor discussions in social media groups. Their main goal is to interact with the ministry online and create a welcoming digital environment.

As part of the volunteer recruitment process, taking the necessary precautions to screen new volunteers and ensure they're the right (and safest) fit for your ministry is vital. Here's the screening process I used in ministry to help ensure volunteers understood their role's importance and expectations.

Volunteers who serve with kids should:

1. Attend for 6+ Months - Serving is an essential step in being part of the body of Christ. Before a volunteer steps into their role, you want to ensure they are getting spiritually fed, connected, and, most importantly, desiring to serve at your church as a part of their commitment to making this church family theirs. Officially becoming a member of the church is a requirement for the lead volunteer positions only. Becoming a member means that they support the church's vision and agree with the theological framework of the church family.

2. Complete a Volunteer Application - All new volunteers must complete a volunteer application. The information helps ministry leaders collect contact information, screen volunteers, and gather references.

3. Pass Background Checks - All new volunteers over 18 must complete and pass a background check as part of the application process. Some top sites children's ministry leaders use to run background checks on volunteers include: First Advantage, Protect My Ministry, Verified Volunteers, Secure Search, Checkr, Ministry Safe, and state police departments.

4. Sign a Volunteer Covenant - All new volunteers must read and sign a Volunteer Covenant that outlines the expectations and confidentiality policies for all volunteers who serve. The covenant includes details about social media and digital policies discussed in the safety and integrity chapter.

You can use digital methods as part of your recruiting and screening process with volunteers. Create an onboarding process or pipeline for all volunteers who serve in ministry. Procedures may include steps like: completing the volunteer application, passing a background check, signing the volunteer covenant, completing training, being shadowed one Sunday, and being ready to serve, etc. Use a simple Excel spreadsheet to track each new volunteer's progression through the onboarding process or explore some of the options available in your volunteer software. Knowing the steps you need volunteers to take during recruitment ensures the safety and preparedness of team members. The digital tools available for this kind of system help streamline the process of recruiting and onboarding volunteers.

Coordinate Volunteers

Once you recruit your team of rock star volunteers, then it's time to schedule them. Coordinating volunteers is a significant part of every children's ministry leader's role, and it can often be one of the most time-consuming tasks. But digital tools and systems like Planning Center Online, SignUpGenius, Slack, or a well-organized Google spreadsheet can help make the task of volunteer coordination more manageable.

Volunteer teams are typically divided into age groups or serving teams, most often based on the age group they serve with. Joana, one children's ministry leader in Nevada, shared in a children's ministry Facebook group how she uses the free version of Slack to organize her teams and communicate with them. All team members are added to channels specific to their area (nursery, preschool, elementary, guest services, etc.) and general channels like prayer requests and random wins. Then, they use the Slack channels for updates, reminders, and allowing volunteers to communicate with each other whenever they will be out and need a substitute.

In addition to organized teams, volunteer coordination also requires the scheduling of volunteers for specific days and times.

Typical volunteer coordination schedules include:

- Volunteers who serve every week - This works exceptionally well for churches with multiple services. Ask volunteers to attend one service and serve during another.

- Volunteers who serve once a month, the same week every month - I used this approach as a children's pastor in a mobile church.

- Volunteers who serve for a whole month, then take a month off, alternating months they serve - This approach works well for consistency throughout a 4-week unit and is my current church's approach.

A weekly detailed volunteer coordination checklist in my children's ministry looked like this:

- Each month, I created the schedule for the month. We were set up with a team of volunteers to serve one week a month, on the same week each month. So we had a team that served on the first week of every month, another team that served on the second week of every month, another team on the third, and so on.

- Every Monday, I sent a reminder email to all the volunteers serving the following Sunday. This email included a reminder about serving, lesson materials for their review, and asked them to confirm they would be there.

- If I didn't hear back from someone by Thursday, I texted or made a phone call to check in and make sure they received the email and would be available to serve.

- On Fridays, an automated email from our scheduling system (we used Planning Center) gave them one more reminder about their role and service time.

I also utilized texting programs like Remind to stay in touch with volunteers, send reminders, and share encouragement. All that coordination becomes possible in the digital world.

This may be showing my naivety and lack of faith in people's commitment to a volunteer role, but I honestly don't know how volunteer coordination happened before the internet. Digital methods of organizing teams streamline the process and frees children's ministry leaders to focus on investing in their volunteers as leaders rather than just making sure they show up when they're supposed to.

The digital world also makes sharing curriculum, files, and lessons with volunteers easier. Most churches have switched to digital curriculum formats, meaning they access their curriculum online rather than receiving a printed book. This doesn't mean their teaching only happens on a video screen or in a digital way. But rather than purchasing physical books with the series lessons, leaders can now access their curriculum online. This allows them to share the lesson with multiple volunteers easily. Some churches provide the lesson on tablets for the volunteers to teach from so they don't have to print anything.

Train Volunteers

Your work with kids is too important *not* to train your volunteer team. Whether it's an annual volunteer training, a weekly or monthly volunteer gathering, or even a few texts or emails with helpful resources and reminders, finding ways to train volunteers consistently is essential. Effective volunteer training should prepare your volunteers with the basics, inspire them with the importance of their role, and equip them to be rock star leaders in their unique position.

Using digital methods to equip your team can help you train your volunteers more effectively and efficiently. Here are a few tips and ideas for tapping into our digital world when it comes to volunteer training:

Use email automation and scheduling to set up newsletters or reminder emails that go out automatically. You can schedule them and not think about them again, yet your volunteers have reminders and resources about their roles.

Make online training easily accessible, like clicking a link in an email or on Facebook, having a dedicated space on your church's website for volunteers, scanning a QR code, etc., rather than asking them to type in a long URL that they won't remember. You may also choose to host a Zoom meeting for your volunteer meeting instead of gathering in person. We even had a children's ministry volunteer page on our website. It was password protected, had information for current volunteers, and hosted a monthly training blog post. I then used that blog post at our monthly meetings and on social media.

Make digital training bite-size. One side effect of our digital world is that attention spans are getting shorter. Whatever your opinion on this, it's a reality you now must navigate. Don't try to host your entire two-hour, in-person training online unless you break it down into sections or units that volunteers can complete one at a time. A short monthly video from you can go a long way.

Sam Fowler, a children's ministry leader in Texas, found that sharing one-to-five-minute training videos with his team each month allowed volunteers to train on their own time and still learn needed inform ation.[34] The videos covered topics like the overall vision of the ministry, specific, detailed tips for leading kids, bathroom procedures, and more.

Use your social media platforms. Facebook Groups allow you to create units within the group that walk members through a series of learning milestones. I've seen some churches creating online courses for their volunteer training, whether through an official online course platform or even by creating a YouTube video series. Consider using Instagram Reels to make funny videos about new policies and procedures to help communicate some of the less exciting, but still important, information.

One of the ways I connected digitally with my volunteers (especially since we were a mobile church and had no permanent space) was through a Volunteer Facebook Group. The Volunteer Facebook Group was a private group, meaning volunteers could request permission to join, but only I (the Children's Ministry Director) could approve their request. All posts in this group were private and only seen by members of the group. Volunteers could request to join the Facebook Group after training. We used this secret group to ask for subs, share news or funny stories from time with our kids, make announcements about upcoming events, celebrate our volunteers of the month, and have fun together (the meme game was strong).

Tapping into digital resources allows you to train volunteers even when you're not with them, showing respect for their busy schedules and lives outside of ministry. So, use digital channels and methods to communicate with your team, keep them in the loop, and equip them to serve in ministry.

Invest in Volunteers

Investing in your team is a key element in leading volunteers; continue pouring into them and appreciating them so they know they are part of the flock, not just another warm body in the room. Just as you foster community with and for the parents in your ministry, do the same with and for volunteers.

Invest in volunteers by knowing about their lives and getting to know them on a personal level. Investing in volunteers also means celebrating them! You can utilize digital ministry strategies to celebrate and therefore retain volunteers in children's ministry.

One really easy, but often overlooked, place to start is to say thanks. As often as possible, thank the volunteers on your team verbally or in writing. Send weekly "thank-you" emails to all who served on Sunday. Think of specific reasons you are thankful for the volunteer, and thank them for who they are, not just the job they fill. For instance, "Thank you for using your creativity to help our preschoolers understand more about the story of Jonah today" instead of "Thank you for teaching the preschool class today." Give volunteers shout-outs on social media and highlight the rock stars. What if you posted on your ministry social media accounts about a different volunteer in your ministry every Thursday, highlighting where and how they serve, and maybe including a quote about what they love about serving?

Give volunteer appreciation gifts. Let your volunteers know you appreciate them through small gifts at least every semester and every quarter when possible. These gifts don't have to be big or expensive and can be as simple as a punny printable taped to a candy bar. For electronic gifts, consider digital gift cards or ecards. Some churches also provide a QR code or barcode for gift cards to local coffee shops. Volunteers can swing by and pick up their favorite drink, and the ministry covers the bill. I've been on the receiving and giving end of this idea, and it's so much fun! It makes people feel special, and we could all use a little pick-me-up of our favorite drink. (Check out the Time to Update Resources page online for steps on how to set up digital QR code gift cards for your ministry.)

Consider hosting a Lead Volunteer gathering. You may only be able to host a gathering for part of the volunteer team, but gather key or lead volunteers at least twice a year for dinner and discussion. Ask for their input and evaluation of the ministry, thank them for their hearts of service, then cast a vision for the coming months. This type of meeting can even happen online via Zoom or another meeting platform, allowing everyone to participate from home. You may consider sharing an online feedback survey with your key leaders before this meeting so you can gather feedback to discuss during your time together.

Celebrate a Volunteer of the Month. Highlight a volunteer from every age group each month as a Volunteer of the Month. Volunteers of the Month can receive a small gift card or ministry t-shirt and a special shout-out on social media. Here are a few ways to earn Volunteer of the Month: always be on time, be prepared for lessons and activities, go the extra mile to make a child feel special (know their names, ask about their week, etc.), respond to scheduling emails in a timely manner, engage in conversations on the training articles or Volunteer Facebook Group posts, etc.

Want to know one of the best ways to retain volunteers? Celebrate successes and focus on the big picture. Share stories of success within the children's ministry with volunteers, whether in person, through email, or on a Volunteer Facebook Group. Tell them about a child who accepts Christ or prays for the first time and remind volunteers how they play a part in the big picture of a child's faith development. Rocking babies helps them feel safe, so when they become preteens with questions and doubts, they already know the church is a safe place to express those questions and doubts. Teaching elementary schoolers the overarching story within the Bible helps them better process how they fit into that story themselves when they are in high school and searching for their place in this world. Giving young kids the opportunity to serve helps them discern their gifts and passions that God can develop as they grow and pursue careers. Understanding this big-picture approach to ministry and sharing how God works in a child's life helps foster faith and a volunteer team that lasts.

From recruitment to training to volunteer appreciation, managing volunteer teams is a vital task leaders must do well to succeed in ministering to the next generation. The digital world helps you recruit, coordinate, train, and invest in your volunteer team easier and in a variety of ways.

Action Steps

1. Determine digital volunteer roles and start praying and planning for God to fill those roles.

2. Evaluate your current volunteer scheduling and coordination process. Is there a digital tool or automation that could help streamline or improve the process? Check out the Bonus Tools and Resources chapter in this book for resources to explore.

3. Choose one form of digital communication (email, short video, text message, etc.) and schedule a small bite-size training for the volunteer team.

4. Schedule a social media post that gives your awesome team of volunteers a shout-out and a big thank you for all they do.

Don't forget to check out the resources page online at deeperkidmin.com/timetoupdateresources for tools and printables to help with your action steps!

Chapter 7

Social Media

MARK 16:15 SAYS, *"He said to them, "Go into all the world and preach the gospel to all creation."* With over 3.6 billion users worldwide, social media allows believers to "go into all the world" in powerful and far-reaching ways. Churches and ministries that use social media well can create new and effective avenues for discipleship and sharing the gospel.

But, the world of social media is ever-evolving and massive, which can overwhelm and intimidate even the most confident leaders. So, here are four ways to implement a social media strategy into your children's ministry without causing a mental breakdown.

1) Be Where Your People Are

I have some good news for you: You don't have to be on every social media platform known to man, and not every social media channel is the right fit for your ministry. So, when choosing what social media platforms to use, be where your people are.

Focus on two groups: the kids you serve and their parents or grown-ups. Most kids don't have social media accounts, and several platforms have a minimum age limit of 13, so technically, legally, kids shouldn't be on social media. But research (and parents) show kids interact with social media at a minimum by watching YouTube. Pew Research has found that "81% of *all* parents with children age 11 or younger let their child watch videos on YouTube, with 34% indicating that they allow their child to do this regularly".[35]

Many children's ministries, especially during the COVID-19 pandemic, launched their own YouTube channels to connect with kids online, sharing gospel messages and offering livestream versions of their regular programming that kids could watch at home.

Trennon Roye, a children's pastor in a small town in Friona, Texas started a YouTube channel for his children's ministry that successfully encouraged kids to dive into God's Word and invited them to be part of the content creation.[36] He did a lot of research upfront, exploring popular children's ministry YouTube channels, and he noticed most of them used adults, not kids, to produce the videos. Trennon knew about popular YouTube channels the kids in his ministry were watching, and nearly all featured kids. He also knew that in this day and age, it's every kid's dream to be a YouTuber, so Trennon decided to create a children's ministry YouTube channel that got kids involved in the making of the videos.

Trennon challenged his kids to memorize at least three scripture passages a month. When they did, kids were scheduled to help film the videos for the children's ministry YouTube channel. He devised a simple one-page outline with a baseline hook, scripture passage, and big idea, but let the kids have creative liberty in the presentation and filming. "They know what their friends are going to watch, so I let them take over and create whenever I could," Trennon said. The videos turned into simple 15- to 20-minute videos where kids got to practice sharing about Jesus with others and were excited to tell their friends that they were on YouTube.

The project didn't require a major production budget either. Trennon set up a simple YouTube studio in his children's ministry space with a green screen, a table, a few chairs, and an old recording camera that the main service no longer used for live streaming. They recorded the videos before their mid-week services when kids were already planning on being at church, and Trennon learned how to do the simple video editing himself.

Trennon's experiment bore lots of fruit for his ministry. Kids began memorizing scripture more intentionally, becoming more involved in the ministry, and gaining confidence in sharing the gospel with others.

Darian, one fifth-grade boy who loved the limelight and wanted to be part of the YouTube channel, prioritized memorizing scripture. Trennon often posted the month's scripture in the children's area for kids to see. Instead of playing Gaga Ball before programming began, Darian would take time to study and internalize the month's scriptures to take advantage of every filming opportunity. By the end of the year, Darian had memorized 42 different passages of Scripture. Through filming the YouTube videos, Darian also had a greater grasp of biblical topics like the 10 Commandments and the fruit of the Spirit. Trennon remembers: "He really got a hold of it because he was able to be involved in helping teach. He felt significant and included, and it stuck with him more." I don't know about you, but I'd count having a fifth-grade boy engaged and invested in Scripture as a huge win!

Trennon also shared about a third grader named Colt, who was a little skittish about the YouTube channel idea at first, but then got his entire family involved. Colt came from a farming family, so there were times during the year when their family couldn't make it to church on the weekends or throughout the week. During those times, Colt asked his dad to record him saying the scripture and send the video to Trennon so Colt could stay on the YouTube channel schedule. "This told me that: 1) he was practicing the scriptures at home, and 2) he was getting his parents involved," Trennon said. Colt brought his parents into the Scripture engagement and inspired his first-grade brother to join in, too.

Check out VictoryNEXTgen YouTube channel and the Epic Kid Show series to see the videos Trennon's children's ministry created. They released roughly one video a week for over a year.

A kid-led YouTube channel that encourages scripture memorization and gospel-sharing may be outside your wheelhouse or capabilities. Still, you can share children's ministry YouTube channels with the kids you serve. Giving them new ideas and options for what content to watch on YouTube can be a great way to use social media to connect with the kids in your ministry.

The second group of people are the parents of the kids in your ministry, not only the kids themselves. So when thinking of your social media strategy, think of how to connect with the parents of your kids and be where they are. I've mentioned Facebook groups several times throughout the book so far because they are a valuable tool for my context when engaging with parents and volunteers. But if your parents aren't on Facebook, Facebook groups aren't going to be effective for you.

If you need help figuring out where to start, ask! Poll the parents in your ministry about their social media use. Whether you ask a handful of parents who fit the majority of your church or all parents, ask the parents in your ministry where they hang out online. You could even make it super simple by giving them a checkbox list of the platforms you're willing to be on and invite them to check all that they use, leaving a box or space for them to write in another platform if the one they use isn't listed.

Or create a Doodle Poll and ask them: What are your top three social media platforms? What is the first app you open when you pick up your phone in the morning? Where do you and your family spend the most screen time? (On what app, what device, etc.)

You could also ask them what time of day they tend to hang out online. When I did this, many moms said they were on social media while waiting in the carpool lane or after the kids went to bed, so that told me that some of the best times to post were in the afternoons and evenings. Most social media platforms also have valuable insights that tell you when most of your followers are online, so you can utilize that information to help you decide when and where to post.

What are your social media options? I'll give you a quick overview of the top four most popular social media platforms at the time of publication, but know that social media platforms are constantly changing. I'll focus on an overview because these platforms continually update, add new features, and remove old ones.

The four foundational social channels you need to know about as a children's ministry leader include:

1. Facebook - As of May 2023, Facebook remains the most-used and engaged-with social platform, with 2.9 billion users worldwide.[3] The largest age group is people ages 25-34, so the younger parents in your ministry are probably on Facebook. Facebook allows you to create a Page for your ministry and specialized groups. Facebook is best for sharing links, posts, more text-heavy information, and lots of information simultaneously. If you choose Facebook, I recommend starting with a Facebook Page for your ministry, learning that, then adding Facebook Groups. What's the difference? A Facebook Page is a public profile that allows your church to introduce your services and activities. A Group is a community-based feature that gathers people with the same interests to discuss topics and share their opinions. So the Page is like the front porch of your house so they can check you out and see what you're about, and a Group is like inviting them into the living room for further discussion and connection. Both Facebook Pages and Facebook Groups allow tremendous opportunities for people to interact with your church, though Facebook

Groups do allow for more interaction and engagement. You may have one Facebook Page for your ministry as a whole, then multiple Facebook Groups based on life stage (like for nursery parents, preschool parents, elementary parents, etc.).

2. YouTube - YouTube is the second most popular social media platform, with 2.5 billion users worldwide.[25] The largest age group on YouTube is people ages 15-25, though more and more kids are spending time on YouTube too. Creating a YouTube channel for your children's ministry allows you to share longer videos, like a Sunday service, with your ministry families, and the privacy settings allow you to choose what videos are public and which are private. With the YouTube Live feature, you can even interact with those watching in real time. You can use your ministry's YouTube channel to share videos you create for your ministry with lessons, short devotions, and more. YouTube is also a great resource to go to find resources to share with families if you choose not to post there directly or have your own channel.

3. Instagram - Instagram is more popular with the younger crowd, with nearly 60% of Instagram users between ages 18 and 29, and it is much more of a visual platform. Notice how on Facebook, people see the text caption first when scrolling, then the picture or image (if there is one). On Instagram, the opposite is true. Scrollers see the image first, then the caption. Instagram relies heavily on hashtags to have content discovered by new people. Hashtags can be an excellent tool for collecting similar pictures, say all the ones related to a particular curriculum series, in one place or for your church members to "submit" a picture by posting on their pages and using a specific hashtag. Instagram has other features like Stories and Reels, as they compete with SnapChat and TikTok. Use

Instagram to share visually appealing pictures and graphics of your ministry, what you're learning, reminders about upcoming events, and fun engagement posts. Remember: with Instagram, it's all about the picture or graphic.

4. Twitter - Twitter is for bite-size pieces of information or quick one-liners since you only have 280 characters available per post. One powerful tool with Twitter is the use of hashtags. You can find out quickly what topics are trending in the world, meaning what many people are talking about, using Twitter. It's also a great place to follow local establishments like schools or businesses to keep up with local events. Use Twitter in your children's ministry by sharing discussion questions, main points from Sunday's lesson, Bible verses, and more. Note: When Twitter changed ownership in October 2022, many of its users decreased their engagement on Twitter because of the changes that followed. In July 2023, just before the publication of this book, Instagram launched Threads, a similar platform to Twitter, and time will tell what this new social media platform will offer.

Use hashtags on Instagram and Twitter to engage with the community and potential new people. These platforms allow you to follow and search by hashtag, meaning people who don't follow you or know of your ministry can see your content within their feeds and search results. Consistently utilize a group of relevant hashtags with every post. Search the most popular hashtags related to a topic you are posting about and use those. Some solid hashtags for your children's ministry include:

- Ministry-specific options like #(children's ministry name), location hashtags, or a hashtag related to a particular series or event

- General hashtags like #jesus #Bible #Biblememory #church

The best place to discover relevant hashtags for Instagram is Instagram itself. Take a few minutes to search hashtags within the app; it will show you the most popular and relevant hashtags around a topic or theme.

Some other popular platforms as of June 2023 include Snapchat, TikTok, WhatsApp, and FaithSpark.

- Snapchat remains a popular platform for teenagers and youth, with users who open the app 30 times per day. Because its features allow sending private photos that disappear after a certain amount of time, there are a lot of boundaries you'd need to set up with Snapchat. Still, it could be a platform to explore for reaching any youth volunteers.

- TikTok continues to grow in popularity (and controversy). Like Snapchat, it has a much younger user base, and several youth ministry teams have begun looking to TikTok to connect with their youth. However, be aware of the controversies surrounding this social media app when exploring its use in your ministry.

- Another platform worth mentioning is Whats App. It's a great way to communicate with kids privately and safely online. The app is perfect for families who may not have access to a computer or the internet. It's a simple, secure, and reliable messaging app that allows you to share videos and text messages with groups of people. Path United, a faith-based, non-profit, after-school organization used WhatsApp to stay connected with the children in their program in mobile home communities during the pandemic. For example, they shared a short video (even a simple one like one community director reading a book to her 2-year-old), and then in the video, they asked kids to text in their answer to a question.

- Faith Spark is a discipleship platform and follow-up app that

> connects kids to their faith community.[4] It was initially designed for camps to use to follow up with kids after leaving camp, but they've made some additions and changes in recent years to make it an excellent digital tool to use in your children's ministry! It includes features like: small group (and large group) communications, live online events, and discipleship and Bible skills training. The chat feature allows you to connect with kids and talk with them throughout the week. You can even share polls to encourage them to engage or use other features like quizzes to ask questions or invite kids to ask questions! Like on Facebook or YouTube, you can stream live online events and interact with kids virtually, in real-time, in a safe and closed environment. You can encourage kids to test and grow their Bible knowledge using the quizzes and other features. You could even create games or quizzes based on each week's lesson. And it's a closed platform with many administrative systems to keep kids and volunteers safe. Safety is a top priority in children's ministry, whether online or in-person. For example, even though kids can chat privately with their leaders, all chats are stored in the backend, and admins can view transcripts anytime. There's also the option to set up time filters and image and language monitoring. As you're exploring social media platforms for connecting with your kids and families, Faith Spark is one to consider!

Before you jump into all these social media platforms, let me ask you a question: How big is your social media team? If you just looked around, sheepishly raised your hand, and squeaked out a "me," then be realistic, and don't try to jump on Facebook, Instagram, Twitter, TikTok, YouTube, and What's App.

No one other than paid social media marketers have that kind of time, especially not children's ministry leaders. You already wear so many hats, and a social media manager is a full-time position in many companies and churches. Let social media be an extension of your ministry, not a burden, or you'll give up within a year.

Remember, choose a platform or two that you can consistently and successfully use and stick to it. Unless the polls from your families indicate otherwise, I'd recommend starting with Facebook, then Instagram. FaithSpark is another excellent place to start but getting families on board may take more time and training since they won't already be on the platform.

Choosing the right social media platform starts with knowing who you want to reach. Remember: you don't have to be everywhere on social media. You just have to be where your people are. Find out where that is and start there.

2) Be Real

Authenticity is essential for social media. You must know your audience and where your people are, but you must also know yourself. Your social media strategy must match your identity as a leader and ministry.

You might feel as if you must have polished and professional-looking videos or pictures, but most people are more interested in the raw and real. In fact, when a video or graphic is too polished, it looks more like an ad or commercial, so people keep scrolling!

Use pictures from your ministry (not stock photos) as often as possible. People want to see themselves and their kids. Invite someone (even a professional photographer) to take pictures of your ministry every Sunday or at least once a month.

My church pays a professional photographer to join us for all three services on a Sunday morning twice a year so she can capture the Sunday morning experience. We can use these pictures for months afterward, and the photography sessions happen in different seasons, so we have colder and warmer weather shots. Or maybe you ask a volunteer once a month to be your photographer for the morning, and

I'd recommend adding "event photographer" to your list of volunteers for every special event you host.

Then use a system like Dropbox, Google Drive, Flickr, or another digital storage platform to organize and store photos. That way, as you promote this year's egg hunt, you can quickly and easily pull up pictures from last year's egg hunt to help you invite people and promote this year's event. Graphic images are OK to use, but the more actual pictures you can post, the better.

Let your social media be more than just a bulletin board. If your social media accounts are just posts of reminders of upcoming events, you miss out on a great opportunity to connect with your people. Change your posting approach to fostering conversation rather than being an announcement monologue.

Use the 80/20 rule: let 80% of what you post be about your people, not you, so things that serve, inspire, encourage, or equip your parents, rather than asking them to attend something or go somewhere. Then 20% of your posts can be service notices, event invites, and other reminders. One way to help with this ratio is to remember six content buckets:

1. Inspirational content, which is like a quote or a Bible verse to encourage families.

2. Conversational content, when you ask a question and get them talking and engaging online.

3. Educational content, for when you want to teach them something, point them to a resource, or share an idea for a faith activity at home.

4. Connection-building content, behind-the-scenes pictures and details, personal stories from your ministry, volunteer appreciation, or something that helps them connect with the goings on of your ministry beyond Sunday morning.

5. Promotional content, like an upcoming event or reminder, even something fun like snapping a picture of half of the room all set up for a special event and saying, "Here's a sneak peek at the fall festival happening this weekend! Can't wait for you to see it in person!"

6. And finally, just for fun content like a silly gif or meme, a would you rather question, a funny joke, or a fun holiday.

Let's talk about that last one for a minute. Fun is a big part of kids ministry, so having fun on your social media accounts is OK, too! Post about fun holidays or jokes. I love posting a ridiculous Dad joke on Father's Day, and I love asking people about their favorite ice cream flavor on National Ice Cream Day, which is always the third Sunday in July in the United States. On Instagram, use features like Reels and stories to share funny videos. When my church moved our services back indoors after being outdoors for a season during the pandemic, our whole staff did a couple of lip-sync Instagram Reels using popular songs that we could apply to our message. Not everything you post must be deeply spiritual. It's OK to have fun on your ministry social media accounts.

Keeping those six content buckets in mind will help you avoid being just a bulletin board of announcements online. Varying the kind of content you share on social media will encourage people to follow, and more importantly, engage and connect. Having a variety of posts, not just reminders about upcoming events, helps you be more authentic online.

3. Be Consistent

When setting up your social media accounts, choose handles or names that match your ministry name whenever possible and are consistent across platforms. Try not to be "RISING Kids" on Facebook and "RKidsMinistry" on Instagram. Setting up your profiles and handles correctly is an essential way to make it easy for families to find you on social media. Think about these questions: Does your handle match your organization's name? Is the word "church" or "ministry" used in your handle or profile name? Are you set up as a business profile so you have access to all the features and insights? Hopefully, you can answer yes to a majority of those questions.

Once your page names or handles are all consistent, be consistent with your brand, too. *Brand* can feel like a funny word to use in a ministry context, but it basically just means the overall look of your ministry online. If you can be consistent with the look of your posts, it will help your people quickly recognize posts from you and give your social media accounts a more cohesive look.

Consider creating a ministry brand guide or a graphics standard guide (free sample and template included on the Time to Update Resources page online at deeperkidmin.com/timetoupdateresources) that defines select fonts, feel, colors, and logo for your ministry.

- Fonts - What fonts will you use in your posts and content online? Choose fonts for headings, subheadings, and paragraphs. It's usually best to have no more than three fonts: a classic serif, a modern sans serif, and occasionally a script. Take note that many kids (and even youth) cannot read script fonts easily, so I usually recommend avoiding them in children's ministry.

- Feel - What is the tone of your posts and content? Informal and friendly? Formal and professional? Or something else? Determine the tone or feel that fits your ministry and use it consistently.

- Colors - What colors will you use in your ministry postings? These colors can often come from your logo. A good rule of thumb is to select two main colors and one or two accent colors. Other colors can be used in posts, but having colors to use repeatedly helps with consistency.

- Your Logo - How big is it? Are there different variations? What colors are approved? Put your ministry logo on your posts, in the corners as a watermark, so if it gets shared, people know where it came from, and it can be traced back to your ministry.

A graphics standards guide helps you stay consistent and helps with creation. You don't have to spend hours choosing the perfect font if a majority of posts are a variation of two or three fonts. The graphics standards guide gives you boundaries to help you stay on course and on task so you can create quality graphics without wasting two hours trying to find the perfect corresponding color.

A graphics standards guide is also helpful when expanding your social media team to include volunteers. Even if multiple people take pictures or create graphics, you can still have a cohesive look if everyone uses the same fonts, colors, etc.

One other note about being consistent: You don't have to post frequently, just consistently. If you're posting one or two times a week, that's great! Keep it up! If you're posting one or two times a day, that's great! Keep it up! Aim to post with the same consistency over time no matter how often you post.

If you are just getting started, focus on one or two social media accounts (maybe Facebook and Instagram). Start with two to four posts a week on Facebook and one to three posts a week on Instagram, then build up from there.

When I served as the social media coordinator for our church, we posted two or three times daily on Facebook and Instagram, but we worked up to that. The biggest thing is that you don't want to post four or five times a day for weeks and then suddenly only post once a week. Pick a consistency that works for you and stick with it.

As you decide when to post, use the tools in your business accounts to help. You can look up when the people who follow or like your page are online on Facebook and Instagram. At our church, between 4-6 PM is when the highest number of our followers are on Facebook, and more people are online on Monday than on other days of the week, though only marginally.

An easy way to stay consistent is to create a posting schedule for your ministry. As you experiment with consistency, don't be afraid to try different days and times. Trial and error are a great way to find what works, especially on social media, since it's a world that's constantly changing. Finding consistent posting habits can help your social media ministry accounts reach your people.

4) Be Present

Having a solid plan for content that goes beyond bulletin board announcements is a great place to start, but you must be present on your social media accounts, too.

It's called "social" media because its original purpose was to help people be social online. Remember the first word: social. The point isn't for people to stand back and applaud, hit the like button, or even share a post with a friend; the point is to get people interacting with your ministry and each other and to get into a relationship with the people seeing your posts.

So who can you engage with on your social media accounts? You can engage with the parents in your ministry. You could create Facebook groups for the parents in your church or even smaller segments of your ministry, like moms of preschoolers, moms of preteens, or moms of boys. Use social media to engage with parents by asking questions, sharing ideas and resources, and sharing encouragement. As a millennial mom, I follow many Instagram accounts that offer parenting tips on various topics, from sleep to discipline to spiritual parenting. These accounts probably have fantastic websites, but I've never seen them.

However, I do check out the ideas, tips, and encouragement they post on Instagram daily.

You can also engage with your volunteers. Social media is a great way to connect with your team for encouragement, reminders, and training.

While kids aren't on several social media platforms, you can still connect with them online via kid-friendly or parent-monitored platforms like YouTube, WhatsApp, Zoom, and Faith Spark.

Focus less on more people seeing your posts and more on having interactive conversations with the right people. Engage, and engage often. Respond to comments or other people's posts on your ministry account. If someone posts a comment on your summer event saying how much fun they had, ask a question: "Hey, we're so glad you joined us today! What was your favorite part?" Ask and answer questions whenever possible to facilitate an ongoing conversation.

Social media platforms rate interactive comments and shares at a higher value than likes. There is a hierarchy in terms of engagement, which means your post will organically show up to more people if people interact with your posts through comments and shares.

And, of course, follow back. Periodically review your followers list and identify a few to follow back. By creating this two-way relationship on social media, you are also increasing your visibility to potential new people who may be interested in your ministry.

Engage with other people. Engagement is effortless on Instagram. Scroll through your ministry's Instagram feed, commenting on the posts from the people in your church or from local businesses. Comments help them know you're thinking about them throughout the week, and your comment will show up to all their friends and followers too, allowing you to reach new people.

Another great way to engage online is through live video. Interacting live provides tremendous value. Even if you pre-record the footage and set it as a premiere on Facebook, it shows up "live," and you can interact live in the comments.

And if you want your ministry families to know social media accounts exist, tell them! Integrate social media channels and reminders about them into your weekly communication with families, whether in your family email newsletter, a reminder on a parent take-home sheet, or in passing conversations in the check-in line. Share about it during outreach events or on special invitations. If you want your families and guests to interact with you on social media, you must make it easy for people to find you online.

One aspect of social media is that it gives people access to your ministry 24/7. Accessibility is both a value and a pitfall. Hear me say this, leader: engaging on your social media ministry accounts does not mean you are always available for your families online. I'd recommend checking in at least once daily, but don't feel like you have to be glued to your phone, checking for notifications that come through. Sometimes, like during a live or premiere video, you need to be present and engaged online, but it's OK to set up boundaries for yourself and your social media account ministry use.

On the ministry Facebook page, you can set up automatic replies so that if someone messages you, they get an automated response that says something like "Thanks for reaching out. We'll get back to you as soon as we can. In case of an immediate need, email our children's pastor at (your email)." It's like setting up a digital voicemail.

You can also set downtimes on your phone. These are specific times set when your social media apps are "asleep" or inactive. Everything continues to run behind the scenes, but your phone reminds you to take a break, step away, and focus on something else.

It's wise to schedule your posts beforehand (more on that soon), but make sure you also interact with people as they interact with your ministry online. Just as you show people you're listening in real life through eye contact and body language, show them you're listening online by engaging with their posts and comments. Being present on social media means you are responsive to the relationships forming through online conversations.

Let's Recap

1. Be where your people are.
2. Be real.
3. Be consistent.
4. Be present.

When you focus your time and energy on the social media platforms that your families already use and work to strategically utilize these tools for connection through authentic, consistent engagement, you'll be more prepared for the massive mission field of the social media world.

Creating Content for Social Media

There's no getting around it; creating quality content takes time and energy, but there are ways you can maximize your productivity without losing quality. First, know about the different posts you can make on your platforms.

- Photos, Images, and Graphics – Photos help people visualize your message. Pictures and graphics are critical on Instagram, and they work well on Facebook, too.
- Video – You could post a video announcement, ministry advertisement, interview, funny Instagram Reel, or your Sunday lesson. Videos work well on all platforms, as long as you're aware of what orientation of video works best.
- Links – If you want people to take a next step because of your post, consider including a link. Links can take people to an

article, registration page, website, or another social platform, like a YouTube video. Provide the link to guide people to what you want them to do next. Just be aware of how your link shows up within the platform. For example, Instagram doesn't provide clickable links within a post's caption, but you can include clickable links in your profile.

- Questions and Polls – You can gather information from a poll to help you plan an event or start a conversation by having people post their answers. Instagram stories and Facebook both have great options for polling your people.

- Events – On Facebook, you can set up events, invitations, and reminders, and your people can RSVP online. This feature provides the same functionality as services like Evite, letting you see who's coming to the event, and allowing people share the event on Facebook.

- Facebook or Instagram ads – My church has found success with Facebook ads for some of our special events or small group communities, but that is next-level social media use, and we're focusing on the foundation in this chapter. Therefore, I won't go into detail here, but it may be worthwhile to look into.

You don't have to use all these content options. If you're just starting out, focus on graphics or photos first. I'm merely showing you an idea of the variety of posting types that are readily available.

As you consider what type of post to make, keep the platform you're using in mind. One question I often hear when it comes to social media use: Can you post the same thing on Facebook and Instagram? Utilizing the same post is called cross-posting. The answer is yes and no. Let me explain.

Marketing gurus will tell you to tailor your content to each specific platform and don't cross-post. But let's be honest. You don't have time for that, and that approach is for people whose only job is social media marketing. So it's OK to post the same content on multiple platforms, but you may need to modify it a bit.

For example, if you share an invitation to your VBS event and want people to click the link to register on the Facebook post, you can include the direct, clickable link. However, on Instagram, you'd need to adjust the wording to say, "Click the link in our bio to register", since Instagram doesn't allow clickable links in captions.

If you want to create one graphic that works on most platforms, a square image works best for Instagram and Facebook posts. If you start wanting to share things to Stories, you'll have to modify the sizing.

If you're posting on Instagram Stories or creating Instagram Reels, that takes a very different shape and size than anything else, so be mindful that you may need to modify some graphics or pictures for use in different places.

Next, think about the type of things you want to post about. Start jotting down a list of things you could potentially post about using the six content buckets discussed in this chapter's "Be Real" section.

Ideas for content include:

- Sunday lesson recaps with quotes or main ideas from the lesson.

- Ministry events.

- Ask people how you can pray for them.

- Volunteer shout-outs and appreciation posts. (This also encourages new volunteers to plug in!)

- Any important church-wide information.

- Encouraging scripture.

- Links to resources for parents.
- Pictures from Sunday morning and special events.
- Mission-focused posts.
- Responses to current events or the daily lives of families. Things like back-to-school, spring break, and other events that help show a connection to the daily lives of your families.

You also want to think through what *not* to share. Hopefully, you've outlined this in your social media policy, but here are a few topics I recommend staying away from on social media if you want your ministry to be a positive presence online:

- Making too many announcements for the same event.
- Flooding your timeline with "Come to our church" announcements.
- Excessive content relating to a small number of members (like only nursery parents).
- Repetitive news and information.
- Outdated information.
- Anything that consistently promotes a specific leader based on personality as this can create a culture around a person, rather than the ministry.

Finally, create a schedule for your posts and use tools to help you schedule posts ahead of time. The schedule may look like the same type of post on the same day each week:

- Monday - Share a picture from Sunday with a recap about what you learned or the main takeaway.

- Tuesday - Tip Tuesday is when you share a parenting or discipleship tip.
- Wednesday Wisdom - Share an inspirational quote or Bible verse.
- Thursday - Promote an upcoming event.
- Friday - Sneak peek or prep for Sunday's upcoming lesson (so the learning can start at home).
- Saturday - Post a fun, silly "would you rather" question.
- Sunday - Share an action or application step for parents and kids to take the lesson home.
- Make sure you check out the Time to Update Resources page online at deeperkidmin.com/timetoupdateresources for a social media planner template that can help you get started with planning your content.

Scheduling posts will help with that 80/20 rule too, making sure that 80% of posts are all about people, and 20% are about you asking them to do something. If you plan out your posts, you have a better picture of what you're posting. You can even color code things or put symbols next to posts to see how many announcements vs. engagement you post.

A social media calendar details when, what, and where you post. It should align with the overall church calendar. By having a calendar, you can plan through what major events are happening in the church and make sure you're building awareness about those events on the church's social media channels.

You can still post spur-of-the-moment or unplanned posts, and I'm not saying you schedule things a month in advance and then can't post anything else. But having a schedule for the recurring items and the things you know will happen frees you to focus on other aspects of ministry.

As you're scheduling your posts ahead of time, I recommend using a tool that helps you organize all your weekly or monthly posts in one sitting. They automatically post when scheduled without you having to touch them. I like using Later, but I've also used Buffer, Sprout Social, Hootsuite, and many more. You can also use the free scheduler on Facebook and Instagram to schedule your posts.

When creating and scheduling your social media content, do it in batches, meaning choose one morning a month, sit down, and knock out the posts for the whole month in one sitting. Then schedule things and move on to other tasks for your ministry. Doing larger chunks of content creation and scheduling in one sitting will help you tackle more and stay more organized with ministry social media content.

If you're still feeling burnt out or need help deciding what to post, follow other children's ministries on social media! See what they're posting and adapt those ideas. One of my responsibilities with our church was the social media manager. Whenever I saw a post I liked from another church, ministry, or business, I screenshot and emailed it to myself as a reminder to use it as inspiration for something on our church's social media feed. As you learn from other ministries, a word of caution: make sure you don't get caught up in comparing your church to a church with a different set of resources than you.

When you create engaging content and are real, consistent, present, and where your people are, you'll utilize social media to its full extent and capabilities. Despite its downfalls, social media is a powerful tool for connecting with the people in your ministry through digital means.

Action Steps

1. Send out a quick poll to the parents in your ministry, asking them what social media platforms they use the most. Then research the insights of current social media accounts to see when people are online.

2. Go check out FaithSpark and dream about how you could use it in your ministry.

3. Create or evaluate the consistency of your posts using the graphics standards guide on the Time to Update Resources page online at deeperkidmin.com/timetoupdateresources.

4. Set an 8-minute timer and spend that time interacting with other people online, all from your ministry's social media account (rather than your personal account).

5. Using the template provided on the Time to Update resources page online, develop a social media schedule and plan some new posts for the coming weeks.

Want to dive deeper into your social media strategy for children's ministry?

Scan the QR code, add the course to your cart, then use coupon code *timetoupdatesocial* during checkout for 50% off our Ultimate Guide to Building a Social Media Strategy for Your KidMin online training!

SCAN ME

Chapter 8

Technology in Administration

BE HONEST. DID YOU just read the title of this chapter and groan? "Administration? Ugh. The boring stuff." Or did you read it, squeal in delight, and pull out your color-coded label maker? Either way, stay with me, friend! Whether administration is your top spiritual gift or your worst nightmare, it's a vital part of ministry you can't ignore, and the digital world makes it easier than ever before.

What is Administration?

Administration provides a solid framework for ministry. Without it, things crumble. Ministry is about people rather than programs, and administration allows leaders to serve people more efficiently. It's the behind-the-scenes organization that helps everything run smoothly. It enables you to set standards for how you want to operate as a ministry and creates the organizational structure and culture within your ministry. Ultimately, administration allows you to be a good steward of the time, energy, and resources God has given you.

Many people may view administration as tedious, monotonous tasks that limit their creativity and imagination. And while some tasks can be boring (looking at you, budgets), I encourage you to think about administration like a trellis for your ministry vine. A trellis gives a growing vine the structure and stability it needs to thrive. In the same way, solid administration provides the structure and stability your ministry needs to thrive.

And there's excellent news: technology and the digital world can make the administrative part of your job as a children's ministry leader easier! The digital world's tools, apps, and functions can help strengthen and organize the administrative trellis for your ministry vine so that it takes less work yet still provides the same solid foundation.

Administration covers most of the logistics and "business" side of running a ministry. Incorporating digital tools into this aspect of ministry helps you be a better leader, provides a more organized logistical ministry structure, and can help you connect and communicate with the families you serve.

Leadership Administration

Leadership Administration covers your big-picture tasks and overall time management as the children's ministry leader. Leadership administration includes establishing a vision for your ministry, tracking goals and growth, planning the calendar year, personal time management, networking, and transition seasons.

Establish a vision. The vision statement of your ministry is the driving force of everything you do and builds the foundation for the rest of your administrative tasks. Remember, your vision acts like a colander or pasta strainer for ministry, allowing you to run all new ideas, events, and activities through that filter.

Use the digital world to help you define and communicate the vision of your ministry. If you don't have a ministry vision statement already, use a Google Doc to brainstorm ideas, phrases, comments, etc. and invite other key leaders in your ministry to contribute to the list. If you already have a ministry vision statement, evaluate the vision and how well you carry it out. Utilize writing tools and online thesauruses to find the right words for your vision statement. You could host a Zoom meeting with key leaders to discuss and write your vision statement.

Once you establish the vision, use the digital world to communicate it! Use a graphics program like Canva or Adobe Illustrator to create posters, email headers, and website images with your vision statement. Post the vision statement on your ministry's website to communicate with volunteers and families. Schedule regular social media posts to communicate your vision or share pictures highlighting the values of your ministry. Using the digital world to create and establish a vision for your ministry allows you to communicate that vision on all fronts.

Track goals. Your vision helps you set goals for the year ahead, and your senior leadership probably asks you to develop and share goals annually. But, how often do you think about those goals throughout the year as you're in the weeds of ministry and living Sunday to Sunday? Once you've recorded your three to five big goals for the year ahead, don't stop there. Identify three to five small action steps needed to make those goals happen and record those, too.

Use digital tools to keep your goals front and center on your computer and track progress throughout the year. Asana, GoalsonTrack, and Strides are three solid goal-tracking tools that can help you set and track your progress toward ministry goals. Set reminders in your calendar to look at your goals or even schedule monthly meetings with yourself to review progress. The digital world provides excellent resources for helping stay focused on your goals, and when you take advantage of those tools, you can see more significant development and growth.

Track growth. Though discipleship is not about the number of children you see each week, tracking attendance numbers still has value. Whether you use an attendance system like Planning Center or a simple database set up in a Google Spreadsheet, knowing who attends your ministry week in and week out allows you to plan and prepare for growth.

Many apps or programs also have easy-to-use reports to help you quickly see attendance over time. Some even allow you to set up reminders or notifications if a family has not shown up in over a month, allowing you to check in with them so no one slips through the cracks. Discipleship can only occur when families are connected to your ministry in some way or another, so having an organized plan for tracking attendance and growth allows you to keep up with the families in your ministry.

Plan the calendar year. As you plan the calendar year with special events, special Sundays, VBS, camps, and more, you need one simple but vital tool: an online calendar your entire church staff has access to. Whether it's a Google Calendar, an Outlook calendar, or something else, having a digital calendar that all staff members can access is vital for ministry in today's digital world. It allows you to see dates from other ministries, add your dates to the calendar, and even be aware of church-wide and community events.

Remember, the children's ministry is not the only ministry in the church. Even knowing the major youth group dates can help so you don't schedule a babysitting night the same weekend all youth are on a youth retreat. A shared staff digital calendar ensures that no significant overlaps happen with big events and can even work as a tool for reserving space or rooms within your church.

A personal digital calendar can also help you add planning checkpoints or due dates for each big event. For example, if VBS happens in the first week of June, maybe you purchase final supplies and organize final donations the last week of May, open registration eight weeks before in April, start promoting your VBS in March, communicate with your team in February, and finalize VBS curriculum by the end of January. Setting reminders and notifications in your calendar allows you to spread out the tasks and ensure nothing gets put off until the last minute or forgotten.

Another essential element of planning ahead is looking back, and a shared Google Drive or shared Dropbox drive allows your church staff to see notes from previous events and even share photos of the events. When I worked on staff at my church in an administrative role, one of my jobs was to take notes on each major event, record them, and store them in the staff shared drive so that when it came time to plan the next newcomer dinner, we could quickly and easily review the details of the previous event, including catering details, arrival times, and notes on what worked well or what we'd want to change.

Another part of my role was taking and organizing any pictures or footage from the event, again saving them in the staff shared drive, so that if we needed a picture of the layout for the Ash Wednesday service or wanted to use a picture for the following year's promotion, it was easy to find. A shared calendar and an organized online drive that all staff members have access to provide quick and easy tools for planning out your ministry year.

Establish personal time management. Personal time management leads to ministry time management. You must manage your own time well if you want to manage your ministry's time well. As much as I wish it were endless, time is limited. And your time as a kidmin leader is limited, too. So, how can you make the most of it and use digital tools to make your life easier?

First, plan your time. So much time gets wasted from not having a plan for your time. Even if it's just a short list on a digital sticky note or an email to yourself, at the end of each day, make a list of what you want to get done tomorrow. That way, you only spend a little time up front figuring out what to do. You can jump right in, and it helps your brain focus on what needs to get done instead of where to start. Use a digital to-do list like Asana or a project management tool like Trello to help you stay on track. Most apps have computer and mobile options, so you can access your task lists and goals wherever.

Second, track your time. If you want to use your time more wisely, you have to know where your time goes. Your church may ask you to track your time or hours anyway. Mine didn't, but I liked using Toggl (toggl.com) to keep track of my time. I categorized the tasks I spent time on (Sunday prep, special events, meetings, etc.) and how many hours I worked.

Tracking time helped me analyze where my time was going and work to balance it better. For example, if my time tracker showed I spent lots of time in meetings one week, I knew it was OK to cut down on meetings the next week. Tracking time also allowed me to give myself permission to stop working at the end of the day. My to-do list may have had items remaining, but when the timer showed I had been productive for 8-9 hours, I felt better about stopping and pushing those tasks to tomorrow. (The trick is making sure you're truly productive during the time you track!)

Finally, protect your time. Protect the time off you have and time with your family. They are your first ministry before your role in kidmin, and you must protect your time with them, even if it means you have to say no to something. (And it's OK to say no to something.) At one point in my ministry career, I had to schedule meetings with myself on the calendar so I would take the time I needed to rest and recover. If someone asked to talk or meet with me during that time, I could say I had a meeting and wasn't available. Doing this helped me remember to take time to refill and refuel.

The digital world allows leaders to stay connected to others in extraordinary ways, but sometimes you need to disconnect from the digital world so you can be present in your physical surroundings. Set boundaries with your digital tools to help protect your time. Hear this, kidmin leader: you are not obligated to be on call for the families in your ministry 24/7, and rest is productive. God took time to rest and so should you.

Network with other leaders. Ministry is hard, and it can be lonely. Healthy leaders know they need a network of fellow leaders to support, encourage, grieve, and celebrate with them. One of the best things about the digital world is its ability to connect you to others despite physical proximity. That means you can connect with other children's ministry leaders online through social media groups, conferences, and coaching groups. Take advantage of these online communities of leaders to find out what's working for others, ask questions, and learn new tools and techniques.

Prepare for seasons of transition. If you got hit by a bus today, would your staff or team have access to the documents and apps they need to keep your ministry running? It sounds morbid, but a ministry coach once encouraged me to create a "hit by a bus" plan for my ministry and ensure all vital documents (curriculum, policies, procedures, etc.) were easily accessible by someone other than me.

Setting up folders in a staff shared drive online or in the cloud gives your church access to these important documents in the case of an emergency or if you feel called to move on from your current church. When I transitioned off-staff from my previous church, I intentionally organized the files and folders in the staff shared drive to set up the next leader for success. If you have an emergency or feel called to move on, be a blessing to your church and take advantage of how easy it is to store and share important documents online.

Ministry Logistics

Ah, ministry logistics. These are the tasks most people think of when they think of administration, and yes, they can be tedious. But, with the digital tools available to leaders today, logistical tasks can be more streamlined, efficient, and time-saving. From registration to communication, incorporate technology into your ministry logistics to help you better disciple the next generation of kids and families.

Policies and procedures - Use a word processing tool to write and organize your policies and procedures. Store copies of your policies and procedures document online where your team and families can access them, like in a shared drive, a social media group, and even on your ministry webpage. Communicate about your policies and procedures digitally through email or other means.

Check-In and Check-Out - Use a digital check-in system on Sunday mornings or for your midweek programs to help you track attendance and print name tags. Using technology instead of handwritten name tags allows you to better track who attends, and you don't have to worry about reading someone's handwriting. You can pull up instant reports from the day or a specific time period.

Event Registration - Gone are the days of asking families to turn in a physical registration form and the headaches that can come from paper forms. Using technology for event registration streamlines the process for families and you as the leader. With online forms like JotForm, Planning Center Registration, or even Google Forms, families can sign up for your event from wherever and whenever. Families can often register their child, pay the fee, and download important event details in one spot. Most online registration tools also provide quick and easy lists for the event organizer to see numbers, ages, etc. If you still want to hand out physical papers or reminders for an event, add a QR code on the flier encouraging families to sign up online.

Budgets - Confession: budgeting is my least favorite part of ministry, and I dislike it even more than when a volunteer calls out last-minute. But it's an important administrative task leaders can't ignore, and I'm thankful for the technology that makes setting and keeping a budget more manageable. A budget protects people (including you), maintains accountability, and allows ministry to happen. Especially in children's ministry, when supplies like markers, glue sticks, and Goldfish are necessities, having a budget for purchasing these supplies allows you to do ministry well.

But submitting and getting approval for your funding is just the first step. You may need to re-evaluate your budget throughout the year depending on changing circumstances, and you can use digital tools to create monthly reporting to keep track of your budget. Whether using accounting software like QuickBooks, using your own Excel sheet, or hiring a staff member to manage the various budgets of the church, you must keep track of your spending and income regularly throughout the year. There is no such thing as a budget fairy to keep track of it all for you (but how awesome would that be!), but digital tools like Excel are the next best option. I created yearly and monthly spreadsheets with designated expense categories to help me stay on top of my ministry budget, and you can download a copy of the budget tracker on the Time to Update Resources page online at deeperkidmin.com/timetoupdateresources. Use your digital calendar to set a reminder to designate one day each month to ensure your budget tracker is current so you stay caught up.

Volunteer Coordination – Utilize scheduling programs like Planning Center or SignUpGenius to coordinate your volunteers. As someone who has done volunteer coordination for the children's ministry and the church as a whole, I am so thankful for digital tools that make it easier to schedule teams! Honestly, I don't know how leaders effectively and efficiently coordinated volunteers without pulling their hair out before technology.

Communication - Working with people requires communication, and the digital world has ramped up the methods and ease of communication twenty-fold. When working with your team, find digital ways to share curriculum and lesson plans with your volunteers or teachers. Use communication programs like Slack to help keep your team or staff connected throughout the week. When I was on staff at a mobile church (meaning we didn't have a building of our own and instead met in a local high school), Slack was our go-to communication tool as a staff. We used it to share families' updates, talk about upcoming events, send reminders for in-person meetings, ask quick questions you usually ask as you walk by someone's office, and even share wins or celebrations.

Technology can now even *write* the communications for several of these logistical administrative tasks for you! When using an AI (Artificial Intelligence) chatbot like ChatGPT or Google's Bard, leaders can cut their email-writing time in half, streamline the research process, and explore creative series titles. While I recommend applying acute discernment in the way leaders utilize AI and understanding its limitations, here are a few ways AI chatbots can improve your ministry communication:

- Website Copy - Chatbots can assist with website copy by generating informative, engaging, and SEO-friendly content. This can help increase website traffic and improve user engagement.

- Social Media Captions - Chatbots can help create social media captions that are catchy and attention-grabbing. Chatbots can help increase social media engagement and drive more traffic to your website.

- Event Promotion - Chatbots can assist with event promotion by generating persuasive and informative copy. They can even write the event descriptions for you that you can use in promoting the event online and in person.

- Email-Writing - Need an email that encourages parents to have faith conversations at home or celebrates new kindergarteners? Chatbots can help draft those emails for you!

- Curriculum Prep - If you write your curriculum, chatbots can help generate compelling titles and overviews for sermon series, Bible studies, or other events. By sharing the theme of a series with the chatbot, it can generate related topics, catchy titles, and even possible Scripture passages or an outline of the series. While I don't recommend allowing AI to write your entire series for you, it can offer a great place to start with a structure and overview.

- Creating Discussion Questions for Small Groups - Share the lesson with the chatbot and ask it to make small group questions for kids. Then use these discussion questions to share with parents at home or ask the questions in your ministry setting.

- Answer Tough Questions - If you work with kids long, you know they have a lot of questions. Chatbots can be programmed to provide concise and accurate answers to tough questions on sensitive topics. Of course, always use discernment when using the answers provided.

- Research - Need to find Bible verses focusing on a particular topic? Want to know about possible resources around parenting special needs kids? Asking a chatbot for help can streamline your research process. Just be aware it often does not provide sources for its research.

- Clean Up Your Writing - Chatbots can improve the quality of your writing by identifying grammar and spelling errors, suggesting revisions, and providing feedback on style and tone.

Be aware that many AI chatbots do not have direct access to real-time information or the ability to browse the internet. Their training typically only goes up until a recent date, sometimes up to two years in the past, which means they can't provide information on events or developments that occurred after that date. Employ the acute discernment discussed in the introduction when determining if or how AI can aid you in ministry administration.

AI chatbots can be powerful tools for leaders in a variety of tasks. Leaders can improve their communications and organizational performance by leveraging AI chatbots' capabilities. The possibilities are endless for communicating effectively in our digital world, and smart leaders will use the tools technology makes available.

No Excuse

Ultimately, administration is a fundamental part of leading a children's ministry, and with the help of the digital world, administration is easier now than ever before. With all the apps, tools, programs, and more available to facilitate consistent organization and efficient communication, leaders don't have any excuse for overlooking administrative tasks.

And if administration is not your strength, find an administratively gifted person to help you tackle some of these tasks. They'll be excited to explore the digital tools available for refining systems and procedures in your ministry, and you'll be free to focus on other responsibilities.

Action Steps

1. Check out the Bonus Tools and Ideas chapter for a complete list of tools, apps, and programs that can serve you in children's ministry. Choose one to explore this week.

2. For the next week, use the Time Tracker tool available on the Time to Update resources page online to track your time. Then evaluate where your time is going and what tasks you can delegate, automate, or release.

Don't forget to check out the resources page online at deeperkidmin.com/timetoupdateresources for tools and printables to help with your action steps!

Chapter 9

What's Next?

WE'VE OUTLINED THE BEGINNING steps of developing a ministry strategy and identified seven areas to integrate digital discipleship into your children's ministry strategy. Have you completed each chapter's action steps to help you take steps forward? Here's a quick review of the topics we've covered:

- Safety and Integrity in Digital Discipleship
- *First* First Impression: Your Website
- Teaching Digital Natives
- Partnering with Parents
- Connecting with Volunteers
- Social Media
- Technology in Administration

Now that you have some new tools and ideas for your brainstorming list, what's next? It may not be feasible for your budget, schedule, or abilities to implement all these ideas, and that's OK. Choose one area to work on at a time. You won't (and shouldn't) make all these changes overnight. Start small and choose one way to incorporate the digital world into your children's ministry, then commit to continuous improvement.

So, choose one area or idea, then follow through with these steps to implement digital discipleship in your children's ministry strategy. (Remember, you started with steps one through three in chapter one.)

Step 4: Make a Plan

Now that you've got your dreams list from chapter one and have added to it based on chapters two through eight, it's time to start organizing. Try to arrange your brainstorming list into priorities and length of accomplishment. Divide the tasks into week-long, month-long, quarter-long, year-long, and long-term projects. What tasks would take less than a week to implement? A short-term goal may be something as simple as setting up a social media account for your ministry or ordering new equipment.

What about tasks that may take longer, like a month or a quarter, to complete? A longer goal could be creating Slack communication groups with your volunteer team and educating them about using the tool.

Then, what are the tasks that would take a year or more? These tasks may require more culture change within your leadership and ministry, but rather than labeling them as "unrealistic," place them in the long-term category and give special prayer attention to them, inviting God to do what He does. You're not taking anything off the list; just prioritizing the ideas based on the time needed to implement them.

Now, choose one task or item from each time length, and add a few more details to each one to help turn those ideas into actionable goals. When I think of goals, I think of pirates. Pirates recognize you must do more than merely point the ship in the right direction with a vision statement or great idea. You must do the jobs that make the ship move forward, or it will be tossed around by the waves and end up far off course.

Imagine your children's ministry is a pirate ship, and you are the pirate captain (everyone say, "Arr, matey!"). Setting goals for your ministry helps steer your ship in the direction you want and gives you a map of where you are going. They include the specific tasks and actions that propel the ship forward in the intended direction. Raise anchor! Hoist the sail!

Leaders are usually good at setting goals. But most leaders could be better at actually accomplishing them. How can leaders be more effective at achieving their goals?

Start with prayer. (This really should be a given in our line of work.) Some of my favorite verses I pray when planning are Psalm 20:4, Psalm 33:11, Philippians 3:14, Proverbs 16:3, Proverbs 21:5, and Matthew 21:22. I like to read through and pray these verses before every big planning session to focus my heart and invite God into the planning. Praying also allows God to speak to me about *His* goals for the ministry, so I am not limited by what my brain can come up with.

One of the easiest ways to achieve your goals and make them actionable is to write them down. Studies have shown that when you write down your goals regularly, you are 42% more likely to achieve them, because it engages your brain's creative and logical parts.[39] So, write down your goals on sticky notes, in a journal, or an email.

As you write your goals, there are a few key elements to include that can help them become objectives. Your goals should pass the SMART test and be specific, measurable, achievable, relevant, and time-sensitive. Good goals are time-sensitive, so give each one a deadline, which helps with accountability and follow-through. Good goals are also measurable. You will know that you have achieved your objective because there will be evidence, so think through how you will measure and record success.

Look for the specific steps to help you accomplish each goal. You can only move from point A to point D by going through steps B and C first. And if you were really dreaming big, you might have to go through the whole alphabet from point A to point Z.

Some goals require a change in the church culture, and those kinds of changes don't happen overnight. So, start small with the specifics of your goals and take action steps toward the long-term dream.

Do you recall the ministry funnel or discipleship pathway at the beginning of the book? An easy way to choose which goals to pursue is to ask yourself: How is my ministry serving kids and families in each category or step of that funnel or pathway? Or, who does this specific goal help my ministry better serve?

Remember, the goals of discipleship are depth and growth, not numbers and attendance. So, create steps, not programs or events, as you approach your goals and planning. How does each event or element of your digital discipleship ministry strategy address the needs of one or more categories in the funnel or pathway and help kids and families move into a deeper relationship with Jesus?

Download a Ministry Goal Setting Guide from the Time to Update Resources page online at deeperkidmin.com/timetoupdateresources and use the download to help you turn your ideas into goals.

Step 5: Stay Accountable

Stay accountable by sharing your goals with others and adopt an all-hands-on-deck approach. Now that you've written your goals down, talk them up! Who can you invite on the adventure with you? Will it be a co-worker, your senior pastor, a friend, or even a fellow kidmin leader through Facebook? How and when will you share these digital discipleship goals with them? Text? Email? Singing telegram?

As always, be specific. Then, communicate your goals for the digital discipleship strategy with your team, your volunteers, and the families you serve. Sharing invites others to be part of reaching those goals and gives them ownership within the ministry.

Also, keep your goals visible. I am a big fan of sticky notes. (Some of the notes from when I first started in children's ministry are still above my desk.) So, whether it's a physical place in your office or a digital space on your computer, put your digital discipleship goals where you'll see them often.

The best motivation is positive motivation, so celebrate success. As you reach milestones for your digital discipleship goals, take the time to celebrate! Do your happy dance, share the good news, and celebrate the wins. Celebrating motivates you to keep moving when things get tough and invites people into the great things God is doing in your ministry. Even if the accomplishment seems small, take the time to celebrate it.

Finally, take time to evaluate what works and what doesn't. One positive aspect of the digital world is that it allows you to experiment and adapt quickly based on the results. Be OK with trying something new and revamping things as needed.

Repeat. Repeat. Repeat.

Then, repeat those five steps outlined in chapter one and here with another area or idea: 1) identify your mission, 2) evaluate the current state, 3) brainstorm and dream big, 4) make a plan by setting goals, and 5) track and revise through accountability. As a children's ministry leader, you wear a lot of hats (and have a lot of plates spinning at once), so you won't be able to give your full attention to a robust digital discipleship implementation right away. That's OK. Remember that you probably already use digital ministry strategy more than you realize.

Choose one area you want to improve and start small. My recommendation? Start with your policies and procedures to give you a safe foundation as you incorporate digital discipleship into your ministry strategy. Then work up from there to experiment with new teaching methods, create digital administrative systems, connect with your volunteers, equip parents, explore social media, and more.

And if you need help, ask. Find someone on your team or church to teach you how to use technology. Don't be afraid to ask for help or training in this area.

Look outside the church for inspiration. Talk to some teachers and educators about how they're utilizing technology in their classrooms and if they have any ideas or suggestions that might carry over to your ministry. Do any of your volunteers use technology in their workspaces in a way that might apply in your ministry, too?

If you genuinely don't like technology or don't want to spend time learning it, delegate and pass the ownership and leadership to someone else in this area. The help may come from multiple people or the kids you serve!

I also recommend utilizing tools online to learn. Whether you need a quick answer to a simple video lighting question or want to take a deep dive into learning about social media, there are countless opportunities to learn online. Digital training, YouTube videos, children's ministry Facebook groups, or even a quick Google search about your question or topic will probably bring up a step-by-step video that addresses the concern.

Keep Moving Forward

No matter what happens, keep moving forward. Just like there can be some bugs or errors to work out when you update a piece of technology, there will be bumps, curveballs, and setbacks as you integrate digital discipleship into your children's ministry strategy, but don't let them bring you to a halt. Continue pressing on toward your goal even if you take a different route than expected.

Use the tools God has provided in the digital world to reach the next generation and connect them with each other and Him. David Kinnaman, President of the Barna Group, said in a speech at the Child Discipleship Forum 2022: "Ministry in this digital age is going to be different than in the past. God doesn't change, but the context of doing ministry does." When you intentionally integrate digital discipleship into your overall strategy, you'll find that you not only *update* your methods but also *upgrade* them to be more effective.

I'm rooting for you and praying with you. You've got this, leader, and more importantly, so does God. He's not surprised by the technological advances of our digital world. He's the one who inspired the creativity required for them to exist, after all.

The kids and families you serve are worth the effort to incorporate digital discipleship into your children's ministry strategy. Ministry leaders can leverage the tools of the digital world to engage and equip kids more effectively. For the sake of sharing the gospel with the next generation, it's time to update.

And now you can!

Chapter 10

Bonus Tools & Ideas

USE THIS BONUS CHAPTER as a quick reference guide for when you're looking for a digital tool to help you streamline a particular aspect of your ministry. This list includes all the tools mentioned throughout the book and others I learned about through various interviews and conversations. In May 2021, I asked children's ministry leaders to share apps or digital tools they can't live without in ministry, and I also included those answers in this list. The list was most recently updated at the time of publication, August 2023. Head to the *Time to Update* resources page online at deeperkidmin.com/timetoupdateres ources for direct links to these tools.

Communication Tools:

With Families:

- Remind

- Email
- Facebook Page
- Facebook Group
- Mailchimp or Constant Contact
- Group Me
- Text In Church
- Loom
- Zoom
- LinkTree (for creating a resource page on your Instagram)
- Set up text replacement shortcuts on your phone for quick and easy text messages

With Volunteers

- Remind
- Email/Mailchimp or Constant Contact
- Facebook Page
- Facebook Group
- Group Me
- Loom
- Zoom
- Telegram – Commenter said, "It's an app you can use instead of a group text for communicating. Not all my team has social

media, and group texts are inefficient. You can upload files into it and photos, etc. My team uses it to communicate when calling out last minute to have someone fill their spot on Sunday morning. I also use it to share wins within our ministry and any upcoming events."

- Felt – Commenter said, "Felt is an awesome card making app! You take a photo, load it in, and then type a message and you buy credits. Felt then sends the card to whoever you'd like! I use it to send cards to my volunteers! I'll take a photo of them in their role at church and send a verse with a thank you message!"
- Lead Small App from Orange
- Mailchimp
- Slack

With Team/Staff

- Slack
- Email
- Google chats
- Zoom
- Loom

For Kids/Families:

- Bible App for Kids – (ages 5-9) A great app for kids to read the Bible and interact with Bible stories.

- YouVersion Bible + the Kids Bible Experience – (ages 7-11) Did you know the YouVersion Bible App has a special Kids Bible Experience for preteens? Find it on the home page!

- Minno – (ages 3-12) A video streaming service with faith-based shows for kids. Subscription required.

- Verses – (ages 8+) A Bible memory verse app to help you practice and track your progress.

- Lectio for Families – (ages 5+) A family devotional (reading or audible) based on the ancient practice of lectio divina.

- Parent Cue – (parents) An app designed for parents to help raise kids with faith and character.

- Bible Coloring – (ages 3+) Simple Bible story coloring pages.

- God For Kids – (ages 5-12) This app teaches kids about God and His character. It has 31 interactive, child-centered devotions to spark discussions about God, Jesus, and the Holy Spirit.

- Noah's Bible Memory – (ages 7-12) A Bible memory verse game with 25 different Bible verses to learn through increasingly difficult levels.

- Guardians of Ancora – (ages 8-12) A true game, this app invites kids to collect missing items, explore the world, and learn about people in the Bible along the way.

- SunScool – (ages 5-12) Bible study app for kids that encourages them to memorize scripture, listen to Bible stories read to them, and interact with Bible characters.

- Jesus Loves Me – (ages 2-4) Simple Bible songs, memory

game, and short Scripture videos for the littlest ones in your family.

- Bible Word Puzzle – (ages 8+) Kids read through the Bible and develop language skills by finding words in a random group of letters.

- Play the Bible Word Match – (ages 8+) Kids learn as they play in this game where they must choose the right word to complete each Bible verse.

- Superbook Kids Bible – (ages 7+) A media-rich experience that helps bring the Bible to life with videos and images as well as engaging interactive games.

- Talk About – An email subscription from Awana that equips parents to lead intentional discipleship conversations with their children.

- Collide Kids Podcast - A family-friendly podcast that interviews interesting people about how their faith and life collide!

- Miss Patty Cake's Media Guide - A free downloadable PDF with Miss Patty Cake's top 40 faith-based media suggestions for kids and families.

As always, check these out for yourself before sharing them with your families! Note: Some apps may include ads or in-app purchases. Check out the Time To Update resource page for a printable and sharable list of some of these apps to send to your families!

Note Taking:

- Word

- Google Docs
- Notion – Commenter said, "Basically you can make pages which house all the different notes you need as a children's minister. Then share them with your team, volunteers, pastors, and even parents. It helps corral everything in one place. I use it mostly for my master to-do list and planning."
- EverNote

Planning/To-Do Lists:

- Trello
- Asana
- Monday
- Sortifyd
- ToDoist

Time Tracking:

- Toggl

Scheduling:

- Planning Center
- SignUpGenius
- Calendly (for individual meetings)

Goal Setting/Tracking:

- Asana
- GoalsonTrack
- Strides
- Done (Habit Tracker App)
- Notion

Registration:

- Jotform
- Google Forms
- Planning Center Registration

Check-In Systems/Attendance Tracking:

- Planning Center (People)
- Ministry Tracker
- Roll Call
- KidCheck
- Breeze
- KidEventPro

- Tithe.ly
- Rock RMS
- Lambslist
- KidMinApp (kidminapp.com)

Social Media Scheduling:

- Hootsuite
- Buffer
- Later
- Loomly
- Sprout Social
- Facebook Creator Studio

Graphics Creation:

- Canva (top recommended)
- Adobe suite (takes some learning)
- WordSwag
- LightRoom

Presentation Tools:

- Prezi
- Propresenter
- PowerPoint
- Google slides
- Easy Worship (easyworship.com)
- Media Shout (mediashout.com)

Kids Service Programming

- Planning Center
- Playlister (playlister.com)

Internal Organization

- Google Drive
- Google Calendar
- Outlook Calendar
- Dropbox
- Dashlane
- Totescan – Commenter said, "It's a way to organize our storage area. I purchased the labels from Amazon under $10. The app will scan the labels. You then add items under that label

and can even take photos of it. When I was hired, our storage was a hot mess. I didn't know what we had. After emptying it we organized with this app and labels. Then I got a new coworker. She can easily find whatever we need by searching the app. The challenge is to return things back to the same box!"

Media/Resource Sources:

- Right Now Media
- Pinterest
- Deeper KidMin (deeperkidmin.com)
- Twelve Thirty Kids
- WorshipHouse Kids
- Kidology.org
- KidzMatter.com

Background Checks:

- First Advantage
- Protect My Ministry
- Verified Volunteers
- Secure Search
- Checkr

- Ministry Safe
- Through their State Police Departments

Budgeting:

- Quickbooks Online
- Excel
- Google Sheets

Other:

- Amazon
- Weather Channel App
- Bible App
- Candy Crush (commenter said "It helps keep me sane.")
- Developing a Digital Strategy Online Training from Esther Moreno at https://www.childsheart83.com/estherondemand
- LeadMinistry.com/Lead-Digital
- Zapier (automated workflow)

Acknowledgments

"The only way to do great work is to love what you do and have the support of others who believe in you."
Steve Jobs

I've wanted to be an author since elementary school, but when I first started writing this book, I publicly called it a "project" because the words "I'm writing a book" were too big and scary. After conversations, interviews, and encouragement from the following people, the existence of this book became less intimidating, more realistic, and eventually achievable. So, here's to loving what I do, and more importantly, to those who believe in me:

To Joy Canupp, for asking questions that push me toward action and cheering me on through every season of life, ministry, writing, and business. You are one of the three people who knew about the book from its inception, and you have thrown confetti in celebration every step of the way. Thank you for being a sounding board, a cheerleader, and most of all, a friend.

To Esther Moreno, whose energy and passion for children's ministry leadership exceeds all others. I am thankful for your collaboration, spunk, and friendship. You are an inspiration, sister!

To Josh Denhart, for dragging me out of my comfort zones to pursue goals I never thought possible. Though I was often hesitant to see it, thank you for recognizing the potential in me and caring enough to have the patience to coax it out.

To Sherry Chester, for your patience and grace when editing this book. I promise to pay closer attention to verb tense next time.

To Tina Houser, for sharing suggestions and comments that made me a better writer and creator. Your creativity and encouragement have been invaluable sources of inspiration throughout the years. I want to be like you when I grow up.

To Keith Ferrin, for introducing me to the tools, habits, and systems that made finishing this book possible. Your insights helped me push past the roadblocks in the writing process and finish this book, even at 8 months pregnant.

To all the leaders and friends who allowed me to interview them, including Mark Campbell, Trennon Roye, DeDe Reilly, Brandy Thixton, Jon Torres, Erica Mair, Sam Fowler, and many more. A special thanks to Jessica Bealer for not only allowing me to interview her but also for connecting me with other leaders to interview too. I appreciate you all giving me your time and thoughts to make this book a collaboration of ideas rather than merely a list of my own experiences.

To the staff members at RISING Church and Celebration Church, for extending the grace and space to learn ministry leadership and giving me a playground to experiment with new ideas.

To Katie Hove, my best friend, who asked "How's the book going?" often enough to keep me on track. Our two-hour phone dates are the highlight of my week, and you're the real MVP for talking me through writer's block and for sending cookies to help me celebrate.

Shout out to my parents, Sellers and Martha Carmack, for getting me those finish-the-story workbooks for Christmas as a kid that helped develop my love for writing (even though I'm not writing about what would happen if the gum on my shoe came to life).

A special thanks to my girls, Holloway and Kinsley, for sharing Mommy's attention with this book and for working on your own "projects" alongside me.

And words can't express my gratitude to my husband, Ryan. Your unwavering support, understanding, and sacrifices have given me the freedom to pursue my dreams. Thanks for loving me and putting up with me, even when my mind was so focused on writing that I forgot to close the fridge or turn off the water in the kitchen sink.

Finally, to the leaders and friends who follow Deeper KidMin and helped me launch this book, thanks for joining me on this adventure as we grow deeper together, serving kids in Jesus' name.

About the Author

Brittany Nelson is a former children's pastor and the creator of DeeperKidMin.com. Deeper KidMin is an online hub of downloadable resources made *for* children's ministry leaders *by* children's ministry leaders. She holds a Master of Arts in Ministry degree from Wesley Seminary with a concentration in Children, Youth, and Family Ministry. She works to support children's ministry leaders around the world and volunteers as a 2nd/3rd-grade small group leader at her church. Her other adventures include being a mom to the two sweetest girls, reading as many books as she can, volunteering with her husband's youth group, and drinking lots of herbal tea. She one day hopes to run the Disney Half Marathon dressed as her favorite princess, Belle.

Connect with Brittany:

Email: brittany@deeperkidmin.com

Phone: (678) 524-4096 (But remember, I'm a millennial, so text before you call.) :)

Facebook: @deeperkidmin

Instagram: @deeperkidmin

About the Publisher

DEEPER KIDMIN
growing kids deeper

Deeper KidMin is an online hub of creative resources made FOR children's ministry leaders BY children's ministry leaders that help you grow kids deeper in their relationship with Jesus Christ.

Deeper KidMin serves the kidmin community through two main parts: the blog and the DKM Marketplace.

The blog highlights innovative thoughts and practical resources for KidMin leaders to use in their ministries and for parents to use at home. The blog also includes our Subscriber Freebie Library!

The Deeper KidMin Marketplace is a digital marketplace of resources FOR kidmin leaders BY kidmin leaders. The DKM Marketplace is where children's ministry leaders share the resources they've created for their ministries at an affordable price that won't break your budget. All of our products are instant digital downloads available immediately after payment is received. Explore the DKM Marketplace at deeperkidmin.com/dkm-marketplace/. Want to be a DKM Marketplace Vendor and sell what you've created for your ministry? Find more details on becoming a Vendor at deeperkidmin.com/becomeavendor.

Deeper KidMin has 4 goals when creating and sharing resources:

1. **Grow kids deeper in their faith** – We offer Biblical resources that focus on helping leaders connect with and disciple the kids/families they serve. Ultimately, we want to see kids grow deeper in their faith through the resources we provide.

2. **Offer tried and true resources** – We want to share resources created by people who are in the trenches, leading ministry, in volunteer or paid positions.

3. **Customizable resources for any ministry context** – We all know ministry looks different in different places. We want to provide resources that are customizable and designed for leaders to edit and adapt to best fit their ministry context. One of our goals is to make the leader look like a rock star with the kids and families they serve and position them (not us) as the spiritual leader.

4. **Make a kidmin leader's life easier** – We want to share resources that are easy to use and implement, and we do all we can to make a leader's life easier. That means including word-for-word scripts, hyper-detailed descriptions and instructions, and ready-to-use media elements.

Check us out now at deeperkidmin.com

ONE LAST ACTION STEP

Thank you for reading my book!

I love hearing about what digital discipleship strategies you're pursuing in your ministry.

If you found this book helpful, and especially if you've had success implementing the content in these pages, please take two minutes to leave an honest review:

deeperkidmin.com/timetoupdatereview

1. Friedman, T. (2016). Thank You For Being Late. Farrar, Straus, and Giroux.
2. Hills, Ph.D., T. (2016, December 28). The Evolution of the Written Word. Psychology Today. https://www.psychologytoday.com/us/blog/statistical-life/201612/the-evolution-the-written-word
3. Jones, N. (2020). From Social Media to Social Ministry: A Guide to Digital Discipleship. Zondervan.
4. Internet and social media users in the world 2023. (2023, January). Statista. https://www.statista.com/statistics/617136/digital-population-worldwide/
5. Six Questions About the Future of the Hybrid Church Experience. (2022). Barna. https://shop.barna.com/products/6-questions-about-the-future-of-the-hybrid-church-experience
6. Jon Torres, phone interview on May 6, 2021.
7. Denhart, J. (2023). Lead Digital [online course]. LeadMinistry.com. LeadMinistry.com/Lead-Digital
8. Deluccia, T. (2022, December 23). The Ultimate Guide to Digital Ministry Strategy. The Unstuck Group. https://theunstuckgroup.com/ultimate-guide-digital-ministry-strategy/
9. Scazzero, P. (2016, April 5). Jesus' Upside Down Strategy. Emotionally Healthy Discipleship. https://emotionallyh.wpengine.com/jesus-upside-strategy/
10. Lindley, A. (2022, November 11). How to Do a SWOT Analysis (3 Examples & Free Template). Semrush Blog. https://www.semrush.com/blog/swot-analysis-examples

11. GuardrailInsp_RepairGuide.pdf. (n.d.). Retrieved February 21, 2023, from https://www.scdot.org/business/pdf/GuardrailInsp_RepairGuide.pdf

12. Terms of Service. (2023). YouTube. https://www.youtube.com/static?template=terms

13. Clifford, P. A. (2018, November 21). Can I legally show YouTube videos at church? Twelve:Thirty Media. https://twelvethirty.media/production/can-i-legally-show-youtube-videos-at-church/

14. Mark Campbell, phone interview, December 11, 2021.

15. Rainer, T. S. (n.d.). Seven (Plus One) Deadly Sins of a Church Website. Christian Post. Retrieved February 21, 2023, from https://www.christianpost.com/news/seven-plus-one-deadly-sins-of-a-church-website-96661/

16. Denhart, J. (2023). Lead Digital [online course]. LeadMinistry.com. LeadMinistry.com/Lead-Digital

17. Houser, T. (Ed.). (2018). Home Grown: How to Equip Parents for Spiritual Leadership. Four Rivers Design.

18. Prensky, M. (2001). Digital Natives, Digital Immigrants. On the Horizon, MCB University Press, 9(5). https://www.marcprensky.com/writing/Prensky%20-%20Digital%20Natives,%20Digital%20Immigrants%20-%20Part1.pdf

19. Noonoo, S. (2018, June 26). This Neuroscientist Explains Why Today's Kids Have Different Brains. EdSurge. https://www.edsurge.com/news/2018-06-26-this-neuroscientist-explains-why-today-s-kids-have-different-brains

20. Prensky, M. R. (2010). Teaching Digital Natives: Partnering for Real Learning (1st edition). Corwin.

21. A brief guide to Imaginative Education – IERG. (2018). Centre For Imagination in Research, Culture, and Education. http://ierg.ca/about-us/a-brief-guide-to-imaginative-education/

22. Plass, J. L., Homer, B. D., & Kinzer, C. K. (2015). Foundations of Game-Based Learning. Educational Psychologist, 50(4), 258–283. https://doi.org/10.1080/00461520.2015.1122533

23. Jordan Grissom, personal email communication, January 11, 2023.

24. Jessica Bealer, phone interview on June 22, 2021.

25. Understanding, Teaching, and Reaching Digital Native Students—And Digital Native Caregivers. (2021, March 11). Waterford.Org. https://www.waterford.org/education/reaching-digital-native-students-and-caregivers/

26. Bastian, Karl. (2023, July 6). SOMETHING NEW I AM DOING FOR FOLLOW UP I thought you'd like to see. When a family visits, I text them. [Group Post in Kidology.org - Let's Talk KidMin Facebook Group]. Facebook. https://www.facebook.com/groups/kidologyorg/posts/10160993722206031/

27. Newport, F. (2014, November 10). The New Era of Communication Among Americans. Gallup.Com. https://news.gallup.com/poll/179288/new-era-communication-americans.aspx

28. U.S. e-mail usage by age 2021. (2023, July 7). Statista. https://www.statista.com/statistics/271501/us-email-usage-reach-by-age/

29. DeDe Reilly, personal communication, March 2021.

30. Crouch, Andy. Talk given at Child Discipleship Forum. September 2022.

31. Jessica Bealer, phone interview on June 22, 2021.

32. Deanna Porter, personal communication, 2021.

33. Brandy Thixton, phone interview on June 3, 2021

34. Sam Fowler, phone interview, April 28, 2021.

35. Atske, S. (2018, November 7). Many Turn to YouTube for Children's Content, News, How-To Lessons. Pew Research Center: Internet, Science & Tech. https://www.pewresearch.org/internet/2018/11/07/many-turn-to-youtube-for-childrens-content-news-how-to-lessons/

36. Trennon Roye, phone interview on October 14, 2022.

37. Wong, B. (2023, May 18). Top Social Media Statistics and Trends of 2023. Forbes Advisor. https://www.forbes.com/advisor/business/social-media-statistics/

38. Faith Spark | Digital Discipleship App. (2023). Faith Spark. https://www.faithsparkapp.com

39. Matthews, Gail, "The Impact of Commitment, Accountability, and Written Goals on Goal Achievement" (2007). Psychology | Faculty Presentations. 3. https://scholar.dominican.edu/psychology-faculty-conference-presentations/3

Made in the USA
Las Vegas, NV
23 August 2023

76487177R00111